# Resource Recovery from Municipal Solid Wastes

# Volume II

## Final Processing

Authors

**Luis F. Diaz**
President
Cal Recovery Systems, Inc.
Richmond, California

**George M. Savage**
Vice President
Cal Recovery Systems, Inc.
Richmond, California

**Clarence G. Golueke**
Director of Research and Development
Cal Recovery Systems, Inc.
Richmond, California

**CRC Press, Inc.**
**Boca Raton, Florida**

**Library of Congress Cataloging in Publication Data**
Main entry under title:

Resource recovery from municipal solid wastes.

Bibliography: p.
Includes index.
Contents: v. 1. Primary processing—v. 2. Final processing.
1. Recycling (waste, etc.) 2. Refuse and refuse disposal. I. Diaz, Luis F. II. Savage, George M., 1913- . III. Golueke, Clarence G., 1917-
TD794.5.R455 628.4′45 81-18021
ISBN 0-8493-5613-X (v. 1) AACR2
ISBN 0-8493-5614-8 (v. 2)



Direct all inquiries to CRC Press, 2000 Corporate Blvd. N.W., Boca Raton, Florida, 33431.



International Standard Book Number 0-8493-5613-X (Volume I)
International Standard Book Number 0-8493-5614-8 (Volume II)

Library of Congress Card Number 81-18021
Printed in the United States

## THE AUTHORS

**Dr. Luis F. Diaz** received his B.S. degree in Mechanical Engineering from San Jose State University. He then entered the University of California (Berkeley) where he received his M.S. and Ph.D. degrees in the area of Environmental Engineering. Dr. Diaz was instrumental in developing the solid waste processing facility at the Richmond Field Station of U.C. Berkeley. During the course of developing this facility, Dr. Diaz played a major role in developing the fiber recovery system for extracting cellulose fiber from municipal solid waste. Dr. Diaz then undertook an extensive study of biogasification of various waste fractions. This work dealt with the digestion of organic waste fractions and sludges. This work later expanded to include the study of integrated energy-agro-waste systems. This dealt primarily with the use of waste heat as an energy source in an agricultural complex integrated with a wastewater treatment facility.

Dr. Diaz now serves in the capacity of President of Cal Recovery Systems. He has served as a consultant to various municipalities and governmental agencies, as well as leading the work of Cal Recovery Systems in various international projects. He has given numerous invited lectures at technical meetings and has served as the co-organizer of several seminars. He has also worked as consultant to international agencies such as the United Nations (WHO and UNIDO), the World Bank, the Asian Development Bank, and the U.S. Agency for International Development.

**George Savage** was born in 1948 in Faribault, Minnesota. He received his B.S. and M.S. degree in Mechanical Engineering at the University of California (Berkeley). He currently serves as Vice President of Cal Recovery Systems. He has been actively involved in the development of waste processing technology. He conducted pioneering efforts in the areas of developing test methods and procedures and of testing of waste communition, air classification, screening, and densification systems. He has conducted field testing programs at all of the major resource recovery plants in the U.S. In addition, he has advised equipment manufacturing companies on improvements and devices for recovering various waste fractions from the solid waste stream.

He is a member of several technical societies and has served as a lecturer on various aspects of solid waste management at several national conferences.

**Dr. Clarence G. Golueke, Ph.D.,** has been the Director of Research and Development at Cal Recovery Systems, since 1978. Prior to that date he was a member of the faculty of the Division of Sanitary Engineering of the Department of Civil Engineering, University of California.

Dr. Golueke received his B.A. degree from St. Louis University, Illinois and his M.A. degree from the University of Illinois in Urbana. He received his doctorate of philosophy from the University of California at Berkeley.

He is a member of several professional and technical societies and the honor society of Sigma Xi. He was the recipient of the A.M. Wellington Award (American Society of Civil Engineers) in 1966.

His professional activities include service as consultant to federal, state, and local governments and to industry on various aspects of solid and liquid waste management and environmental control.

Dr. Golueke is the author or co-author of more than 100 publications, a large portion which are concerned with biological systems of resource recovery from solid and liquid wastes. He is also on the boards of five scientific journals.

# RESOURCE RECOVERY FROM MUNICIPAL SOLID WASTES

**Luis F. Diaz, George M. Savage, and Clarence G. Golueke**

**Volume I**
**Primary Processing**



**Volume II**
**Final Processing**

# TABLE OF CONTENTS

Chapter 1

# INCINERATION

## I. INTRODUCTION

### A. Methods of Thermal Energy Recovery

Energy can be recovered directly from municipal solid waste (MSW) as heat, or the waste can be processed into a storable fuel. Direct recovery is accomplished by mass burning. Indirect recovery can take several forms and involve many different procedures, each of which may be fitted into one of the three categories: physical, thermal, and biological. In physical processing, the MSW is processed such that its combustible and noncombustible fractions are separated one from the other. The physical characteristics of the combustible fraction are further altered to enhance its utility as a fuel. The resulting combustible product commonly is termed "refuse-derived fuel" or simply "RDF". In thermal processing, usually termed "pyrolysis", the goal is to convert the waste almost entirely into a combustible gas. However, the usual outcome is a collection of solid, liquid, and gaseous products that are more or less combustible. The product from biological processing may be either gaseous or liquid, depending upon the system used. Descriptions and discussions of biological and thermal processing are presented in detail in other sections of this book.

Of the several thermal energy recovery processes, mass burning and RDF production are regarded as having the greatest potential, even though several obstacles may impede the attainment of the full realization of that potential. Some of the obstacles are technical in nature; others are related to marketing. Both types of obstacles can be overcome in part by using the heat energy to generate steam and then market it (steam) off-site for the generation of electricity, for heating and cooling, or for a combination of the two. However, such an approach is circumscribed by limitations that restrict its application to only a few localities. Consequently, public officials, particularly in large metropolitan areas, are forced to go on to the next potential solution, namely, on-site generation of electricity. A difficulty attending the generation of electricity is that the community or agency must construct a power plant and install transmission and distribution equipment. In short, the community would have to assume the functions of an electrical utility. Two alternatives to steam and electrical generation are (1) sell the fuel (RDF) to individual customers for use in power generation by them; or (2) sell the fuel to the local utility. The first approach could prove to be quite costly. On the other hand, with the second approach, not only would the utility benefit through the use of an inexpensive fuel, the disposal problem would also be significantly lessened. However, it should be noted that even though it would seem that utilities would be ideal users of RDF, certain technical, economical, and institutional issues remain to be resolved before the utilization becomes a common practice.

#### Incineration

The use of incineration as a means of disposing of municipal refuse is by no means a recent one, as is attested by the fact that it was described as a viable and ongoing practice in a book published in 1901.[1] The early interest in incineration stemmed from the fact that in terms of disposal, a maximum volume or weight reduction could be achieved. In the U.S. this early period of promise came to an all but complete end in the mid 1960s, a time when extremely few of the early incinerators were as yet in operation. The responsible problem was mainly one of an excessively large emission of particulates.

However, it should be pointed out that the early popularity continued to be strong in western European nations. A major contributing factor to the continuation of the acceptance in Europe was the practice prevalent there of incorporating energy recovery into the incineration process. A belated recognition of the energy potential of incineration finally is bringing about a resurgence of interest in the practice in the U.S. An important factor in this recognition is the realization of the continuing increase in the heating value of refuse being brought about by a corresponding rise in the ratio of plastics and paper to food preparation wastes and garden debris. In recent years, Japan has taken the lead over Europe and the U.S. in terms of numbers and capacities of incinerators.

## II. TECHNOLOGY

### A. General Design Features

Key features of an incineration facility are (1) the tipping area; (2) the storage pit; (3) the equipment for charging the incinerator (typically, a crane or a front-end loader); (4) the combustion chamber; (5) the stack emission cleaning equipment; and (6) the boiler, if energy is to be recovered.

In a typical incinerator operation, the municipal solid waste (MSW) is discharged from collection vehicles either onto a tipping floor or directly into a storage pit. The pit serves a twofold purpose: (1) It permits the storage of an amount of refuse sufficiently large to ensure, if desired, a 24-hr/day, 7-day/week operation of the incinerator; and (2) it provides an area where large noncombustible items can be removed, and the remaining wastes can be blended into a fairly uniform and constant charge. The waste is transferred from the pit to a charging hopper. The charging hopper is designed to maintain a continuous feed of waste into the furnace. The waste falls from the hopper into the furnace and onto the furnace stoker, where combustion takes place.

The furnace is the essential element of an incineration system. It may be rectangular or cylindrical in shape, and may consist of only one chamber or may have a primary and a secondary chamber. The principal function of the secondary chamber is to provide the conditions needed to complete the combustion process. The size and shape of the furnace usually are determined by the manufacturer, and usually are based upon a number of parameters, among which are solids and gas flow rates, residence time, and bed depth. The temperature in the furnace commonly is maintained at roughly 900 to 1000°C.

A stoker is a series of grates provided with openings through which air can be passed. Generally, grates are movable (vibrating, rocking, and reciprocating). The movement of the grates serves to agitate the refuse and thereby promote combustion, as well as ensure the removal of the residue from the furnace.

Air for combustion is forced into the furnace through and around the grates and through the sides or through the roof. Air forced through and around the grates is known as "underfire air", and air forced through the sides and roof, not surprisingly, bears the designation "overfire air". Overfire air typically is introduced through jets positioned at specific points in the furnace. It is used to regulate combustion gases driven off during the incineration of the refuse.[2] The flow of the air and combustion gases is controlled by forced- and induced-draft fans to provide from 50 to 100% excess combustion air. Air is forced into the furnace by means of forced-draft fans, whereas it is drawn in by means of induced-draft fans. Both types of fans are used in modern combustion units. Central overfire and underfire air is provided by the forced-draft fans, and the flue gases are exhausted by the induced-draft fans.

### B. Classification

Incinerators may be classified on the basis of: (1) the recovery or nonrecovery of energy; (2) the state in which the residue (e.g., slag and ash) emerges from the combustion chamber; and (3) the shape and number of furnaces (e.g., rectangular and multiple). Energy is recovered by way of the introduction of a waste heat boiler for steam generation. Incinerators designed for energy recovery can be classified into several groups. Of the types, the waterwall and the modular incinerators have been the ones most used in the U.S.

## III. WATERWALL INCINERATORS

The application of the waterwall furnace, originally developed for the combustion of low-grade coal, was a response to the need for combustion chambers more efficient than those of earlier refuse incinerators. The earlier incinerators were designed mainly for disposal, and energy was a minor consideration. The inferior design together with the relatively low heating value of refuse generated at that time resulted in a release of thermal energy barely sufficient to complete the combustion of the wastes. On the other hand, the design of the waterwall incinerator is such that combustion efficiency is promoted and energy recovery is facilitated. Moreover, the incinerator can be readily modified to minimize the emission of air pollutants.

The waterwall furnace began to come into use for incinerating urban wastes in the early 1960s. Thereafter, its popularity spread rapidly throughout Europe with the result that in the succeeding years several hundred units were built that were within a size range of 120 to 1600 Mg/day (130 to 1800 TPD). In sharp contrast, only a few waterwall incinerators have been built in the U.S. The popularity of the unit in Europe is due to a number of factors, among which are lack of land suitable for land disposal, government subsidies, availability of markets for steam, and relatively high costs of fossil fuels.

### A. Description

The overall incineration system is essentially the same as that of a "conventional" refractory-lined furance. The main difference in the system described herein is in the furnace design. In this case, the walls of the furnace consist of closely spaced tubes through which water circulates. The water circulating through the walls absorbs thermal energy radiated from the burning wastes. The water cools the furnace walls and simultaneously reduces the temperature of the exhaust gases, thus decreasing the volume of flue gases to be treated. More energy is recovered from the combustion gases in waste heat boilers by producing steam. Waterwall boilers can be designed to produce high pressure, superheated steam. Air pollution control equipment typically used for treating the waste gases includes wet scrubbers and electrostatic precipitators.

### B. Performance

Certain key factors must be considered in an evaluation of the performance of an incinerator, regardless of type. They include weight and volume reduction, composition of flue gas, quantity and composition of residues, and boïler efficiency. With the data on these factors it is possible to determine at what percent of capacity a unit is operating, the temperature and pressure of steam being generated, the quantity of combustible matter present in the residue, and the type and degree of treatment required for the residues prior to final disposal.

Performance characteristics that may be expected of a waterwall furnace are exemplified by those of the waterwall unit in the Nashville, Tennessee incineration facility.[3] Data collected in the operation of the facility indicate that when it was operated at full capacity (325 Mg/day (350 tons/day)), the steam production amounted to 49,000 kg/hr of 2525-kPa/300°C (107,000 lb/hr of 366 psi/573°F) steam. This production was equal to about 98% of the plant's steam producing capacity, and a boiler efficiency of about 72%. Refuse burned in the Nashville incinerator had an "as-fired" heating value of about 11,600 J/g (5000 btu/lb), and was reduced in weight by slightly more than 78%. According to the data on air emissions, the average uncontrolled particulate loading was 3.35 g/Nm$^3$ (1.46 g/dscf) at 12% $CO_2$, and the $NO_x$ concentration in the flue gas, 146 ppm at 9.5% $O_2$.

**C. Problems**

The design of a modern incineration facility has become a very complex undertaking and can pose a number of problems. In particular, there are three potential problems that must be given special attention. These are (1) system availability; (2) fouling of heat transfer surfaces; and (3) corrosion.

The system availability of waterwall furnaces characteristically is only about 80%. The limitation on availability is mainly a result of the need to frequently clean out grates that have become clogged with partially burned or noncombustible materials in the fuel. The frequency of the downtimes necessitates a duplication of equipment (i.e., 100% "standby"). Thus, two 62- or 72-Mg/day waterwall units would be recommended for an operation in which 91 Mg (100 tons) of refuse had to be burned per day. In addition, in order to meet maximum reliability in the case of energy production, the furnace should be capable of firing a fossil fuel.

Almost all of the heating surfaces are subject to fouling. This is caused by the deposition of slag and fly ash. Research efforts, primarily on coal, have demonstrated that ash deposition can be alleviated by modifying the furnace design, by the introduction of additives, and through the improvement and proper use of boiler-cleaning equipment. Among the modifications in design are those geared to the provision of sufficient volume and detention time to achieve complete combustion of all combustible solids and gases, and of enough heat absorption by the furnace to dry the ash before it comes in contact with the boiler tubes. The heat transfer surfaces in convection passes, particularly those in the high-temperature zones, should have wide spaces.

Corrosion is a major technical problem in the operation of refuse-fired boilers. Basically three types of corrosion can take place: (1) corrosion due to the existence of a reducing atmosphere; (2) halogen corrosion; and (3) low-temperature corrosion.

Corrosion caused by a reducing environment usually takes place in mass burning facilities. It is precipitated by the products of partial combustion which are a consequence of a reducing atmosphere. A reducing atmosphere may be a direct result of poor distribution and stratification of air or fuel, or of both. These conditions could promote the formation of carbon monoxide and hydrogen sulfide. The two gases can react to remove the protective coating on the furnace tubes and expose the tubes to additional corrosion.

Halogen corrosion is attributed largely to the combustion of Cl-containing materials (e.g., PVC, NaCl) in the refuse. Although halogen corrosion has been recognized for many years, a certain amount of disagreement exists as to the mechanism involved and to the temperatures at which it takes place.

Points of low-temperature corrosion are those at which the flue gas comes in contact with surfaces that have a temperature lower than the dew point of water vapor. HCl and $SO_x$ dissolve in the water of condensation to form corrosive acids. Dew-point temperatures may be encountered in air pollution control equipment, in air heaters,

**Table 1**
**MANUFACTURERS OF WATERWALL INCINERATORS**

| European manufacturer | U.S. representative/manufacturer |
|---|---|
| Von Roll (Zurich) | Wheelabrator-Frye[a] |
| Josef Martin (Munich) | Universal Oil Products[a] |
| Voklund (West Germany) | Waste Management[a] |
| Voklund (West Germany) | Trans Energy Systems[a] |
| Voklund (West Germany) | Browning—Ferris Industries[a] |
| | Babcox and Wilcox[b] |
| | Combustion Engineering[b] |
| | E. Keeler[b] |
| | Foster Wheeler[b] |
| | Zurn Energy Systems[b] |
| | Riley Stoker[b] |
| | Detroit Stoker[b] |

[a] Market complete systems.
[b] Market components only, e.g., boilers and stokers.

and in economizers. Low-temperature corrosion can, of course, pose a serious problem when the boiler is taken out of service.

### D. Manufacturers

Waterwall combustion systems and components are available from several manufacturers in Europe and in the U.S. In Table 1 is a list of manufacturers in the U.S. and of European manufacturers who have representatives in the U.S.

### E. Types of Marketed Systems

The three types of waterwall combustion systems presently on the market are distinguished from one another on the basis of the form in which the refuse is burned. The form can be one of the following three (1) raw ("as-received") refuse; (2) shredded refuse; and (3) refuse derived fuel (RDF). In the first type, also known as "mass" or "bulk" incineration, the raw or as-received municipal solid waste is dropped onto moving grates which move the refuse through the furnace. Inert material falls off the end of the stoker and is quenched with water. The cooled residue is then stored or transported to a truck for final disposition. In the second system "as-received" MSW is coarsely shredded and then is mechanically or pneumatically introduced into the furnace and incinerated on a moving grate. This method of combustion is termed "semisuspension firing", because the waste is ignited as it falls through the chamber and, hopefully, its combustion is completed on the grate. A more extensive processing of MSW results in the production (i.e., separation) of a light, combustible fraction, which is the refuse-derived fuel or RDF described in another section. RDF can be used directly in a boiler, or it can be used as a supplemental fuel in boilers especially designed to fire coal, wood, or other biomass.

#### *1. Units for Burning Unprocessed Refuse*

Energy recovery from as-received refuse in waterwall incineration has the longest record of application in that it has been used for over 20 years in the U.S. and Europe. Prompted by the more or less successful records of the waterwall combustion systems in Europe, interest in the potential of the system for energy recovery is intensifying in the

U.S. The first large-scale waterwall incinerator in the U.S. was built at the U.S. Naval Station in Norfolk, Virginia. This facility was designed to process 326 Mg/day (360 TPD) of solid waste and has been in operation since 1967. The steam is used for heating and for cooling.

Other facilities that have received a certain amount of attention include a 1084-Mg/day (1200 TPD) plant in Saugus, Masschusetts and a 650-Mg/day (720 TPD) plant in Nashville, Tennessee. A list of other facilities in operation in the U.S. as of 1979 is given in Table 2.

Experiences in the U.S., Europe, and Japan in the utilization of waterwall furnaces for the combustion of as-received solid waste range from excellent to bad. Generally, the instances of unsatisfactory performances can be attributed to faulty design, under capitalization, and poor implementation. The variability of the composition of the incoming waste stream has also been identified as a source of problems.

*2. Units for Burning Shredded Waste*

At the time of this writing, only one plant had been constructed in North America to burn shredded refuse exclusively. The plant is located at Hamilton, Ontario and started operating in early 1972. It was designed to process 544 Mg/day (600 ton/day) of refuse and is processing about 270 to 360 Mg of residential and commercial refuse per day. The Solid Waste Reduction Unit (SWRU) in Hamilton is soon to be joined by another municipal solid waste plant of similar design presently under construction at Albany, New York.

The waste is size reduced so that a part of it can be burned while in suspension. The remainder is combusted on a traveling grate. Overfire feeding of shredded waste should make it possible to maintain a fairly constant temperature within the furnace. Since a fraction of the waste is burned while in suspension, a thin fuel bed is formed on the grate. This condition allows for efficient operation of the furnace without creating a reducing atmosphere near the tubes. The result is the elimination of the need to provide a refractory protection of the lower tubes and a provision for the maintenance of constant steam production.

Size reduction, however, adds another level of complexity to the operation of an incineration facility. Improper selection and operation of the shredders and ancillary equipment (i.e., conveyor, storage bins, etc.) can lead to the development of severe problems. On the other hand, ash handling should be easier due to the small particle sizes. The advantages of shredding the refuse prior to incineration can not be fully evaluated at this time due to lack of data.

*3. Units for Burning RDF*

To date, the main use of RDF has been as a supplement to coal in large power plant boilers. Most power plant boilers are designed to burn pulverized coal in suspension, and hence have the equipment required for handling solid fuels and ash. When RDF is burned, the ratio of RDF to coal typically is such that the RDF accounts for 5 to 20% of the total heating value of the fuel (i.e., coal plus RDF) fired in the boiler. Recently, two large-scale facilities, one at Akron, Ohio and the other at Niagara Falls, New York, have been scheduled to fire RDF only.

Results of projects in which RDF was cofired with coal in large, suspension-fired boilers have brought to light several problems, as yet unresolved, that are related to fuel handling and storage, ash handling, corrosion, and boiler derating. Many of these problems can be attributed to the use of an inferior grade of RDF, i.e., an RDF characterized by a high ash and moisture content, an unpredictable size distribution, and an extremely variable heating value. Factors responsible for the production of such an inferior RDF are described in the section on RDF.

**Table 2**
**WATERWALL COMBUSTION SYSTEMS IN OPERATION IN 1979 IN THE U.S.[a]**

| System | Process | Location | Operating capacity (TPD) | Operating capacity (Mg/day) | Product | Status |
|---|---|---|---|---|---|---|
| Waterwall Combustion | Raw Solid wastes | Braintree, Mass. | 250 | 227 | Steam | Operational |
| | | Chicago, Ill. (N.W.) | 1600 | 1450 | Steam | Operational |
| | | Harrisburg, Pa. | 720 | 653 | Steam | Operational |
| | | Nashville, Tenn. | 400/7 days | 363 | Steam | Operational |
| | | Norfolk, Va. U.S. Naval Stn. | 360 | 326 | Steam | Operational |
| | | Oceanside, N.Y. | 750 | 680 | Steam | Operational |
| | | Portsmouth, Va. | 160 | 145 | Steam | Operational |
| | | Saugus, Mass. | 1200 | 1088 | Steam | Operational |
| | Shredded Solid wastes | Albany, N.Y. | 750 | 680 | Steam | Startup; steam facility in '81 |
| | | East Hamilton, Ontario | 300—400 | 272—363 | Steam | Operational |
| | RDF/suppl. fuel | Ames, Iowa | 200 | 181 | Electricity | Operational |
| | | Columbus, Ohio | 2000 | 1814 | Electricity | Construction |
| | | Detroit, Mich. | 3000 | 2721 | Steam and/or electricity | Construction |
| | | Duluth, Minn. | 400 | 363 | Steam | Startup spring '80 |
| | | Humbolt County, Calif. | 825 | 748 | Electricity | Predesign |
| | | Lakeland, Fla. | 300 | 272 | Electricity | Construction |
| | | Portsmouth, Va. (southeastern Tidewater Energy Project) | 2000 | 1814 | Electricity | Design |
| | RDF/100% | Akron, Ohio | 1000 | 907 | Steam | Startup |
| | | Niagara Falls, N.Y. | 2200 | 1995 | Steam/Electricity | Startup |

[a] The list is modified from the NCRR Bulletin, *Resource Recovery Activities,* 10, 1, 1980.

## IV. MODULAR INCINERATORS

### A. State of the Art

The early versions of modular incinerators generally were batch fed. Moreover, no provisions were made for the control of the combustion air and for the continuous removal of ash. Consequences of these and other design inadequacies were (1) poor combustion performance; (2) slagging in the primary chamber because of the obstruction of underfire air ports; (3) deterioration of refractory material; and (4) a tendency of the ram loader to drag back burning material from the primary combustion chamber to the feed hopper. The problems have been eliminated, or at least ameliorated, in the more recently designed units through the incorporation of such features as automatic charging devices, continuous ash removal, water-cooled underfire air ports, and means for a more careful control of combustion air. Controlled-air package incinerators have been commercially available for over a decade. The modernized versions apparently function satisfactorily, although reported operating experience is not sufficiently extensive to permit a full evaluation.

Facilities to recover energy through the use of modular incinerators are relatively recent introductions to the waste management industry. In keeping with conventional practice, the facilities rely upon the incorporation of a waste heat boiler in which hot flue gases are used in the production of steam or hot water. An idea of the design of a modular incinerator for the production of steam may be gained from the drawing in Figure 1. As yet, no attempt has been made to produce superheated steam, although the possibility of doing so is being contemplated. A list of cities in which refuse is being or will be processed to recover steam is given in Table 3.

Modular incinerators designed for steam production have the desirable feature of being completely flexible in terms of ability to accommodate changes in the waste stream and steam output requirements. The flexibility stems from their being available in 23- to 45-Mg/day (25 to 50 ton/day) modules. As such, they are especially suitable for small communities, as well as for a phased development of energy recovery. Three additional advantageous features are (1) they are factory built; (2) they can be shipped by way of the highway; and (3) they can be assembled in clusters. Although a modular incinerator facility may require frequent maintenance, the maintenance is not difficult, and generally is not sufficiently involved as to necessitate an interruption of the daily operation.

### B. Commercially Available Units

Of the commercially available modular incinerators designed for energy recovery, three were selected for description because they represent the present state of the technology. They are (1) the unit manufactured by Basic Environmental Engineering, Inc.; (2) the Consumat® incinerator; and (3) the Kelly® incinerator.

#### *1. Basic Environmental Engineering, Inc. Incinerator*

Units manufactured by the Basic Environmental Engineering Corporation are designed to treat both municipal and industrial wastes at rates ranging from 33 to 136 Mg/day (36 to 150 ton/day). They can be operated on conjunction with energy recovery units. The incinerator has three chambers, in all of which combustion takes place in the presence of excess air. The primary chamber is of the waterwall type. The temperature within it typically is maintained at about 980°C. In the latest version, the secondary and tertiary chambers have double walls. The incinerator is equipped with two burners designed to burn fossil fuel for start-up, or whenever supplemental heat is needed.

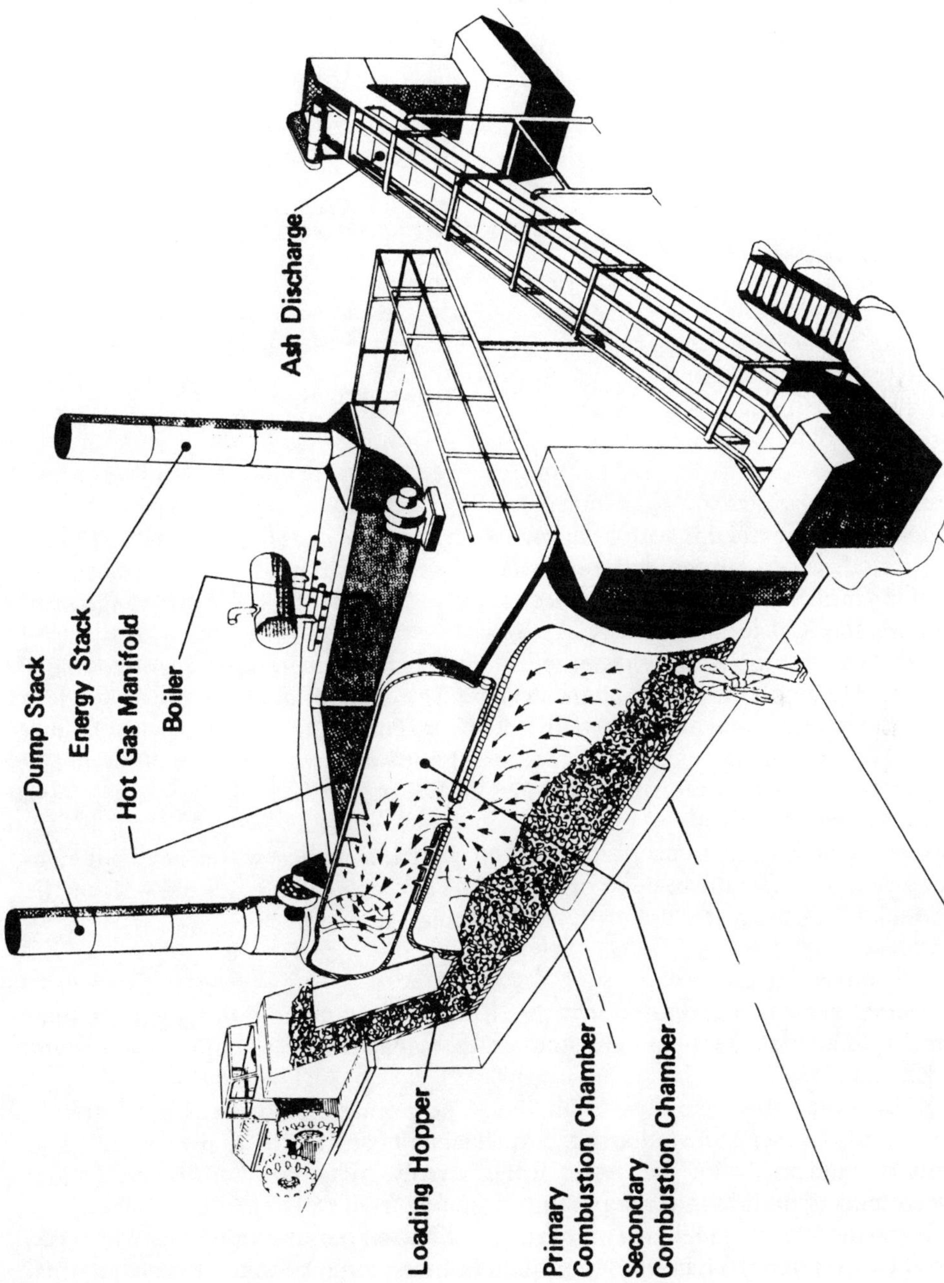


FIGURE 1. Modular waste-to-energy unit.

**Table 3**
**STEAM-PRODUCING MODULAR INCINERATOR FACILITIES**

| Location | Operating capacity (Mg/day) | Operating capacity (tons/day) | Status |
|---|---|---|---|
| Auburn, Maine | 136 | 150 | Construction |
| Blytheville, Ark. | 68 | 75 | Operational |
| Crossville, Tenn. | 54 | 60 | Startup |
| Dyersburg, Tenn. | 91 | 100 | Startup early 1980 |
| Genesee Township, Mich. | 91 | 100 | Operational |
| Groveton, N.H. | 22 | 24 | Operational |
| Lewisburg, Tenn. | 54 | 60 | Construction |
| North Little Rock, Ark. | 91 | 100 | Operational |
| Osceola, Ark. | 43 | 50 | Operational |
| Pittsfield, Mass. | 218 | 240 | Construction |
| Salem, Va. | 91 | 100 | Operational |
| Siloam Springs, Ark. | 14.5 | 16 | Operational |

Modified from National Center for Resource Recovery, Resource recovery activities, *N.C.R.R Bull.*, 10, 1, 1980.

*2. Consumat® Incinerator*

Consumat® elemental units are designed to burn approximately 227 to 900 kg (500 to 2000 lb) of industrial, commercial, or municipal solid waste per hr. The units can be operated in parallel to permit greater throughputs. Typical total capacities range from 18 to 91 Mg/day (20 to 100 ton/day).

The incinerator consists of two chambers. The primary chamber is operated under substoichiometric conditions at about 650°C. The temperature in the secondary chamber is maintained between 980 and 1,060°C by controlling air input into the two chambers. The secondary chamber is equipped with a burner that is automatically activated when temperature drops below 926°C. The burner is designed to use a fossil fuel (natural gas or fuel oil).

The Consumat® system can be used in conjunction with a waste heat boiler to produce steam. Heat for the boiler is provided by hot combustion gases drawn from the incinerator and through the boiler by means of an induction fan.

*3. Kelly® Incinerator*

The incinerators manufactured by the Kelly Company are designed to burn commercial, industrial, and municipal solid waste. Capacities of the units range from 45 to 900 kg/hr.

The Kelly® incinerator has two chambers. The primary chamber is a horizontal, cylindrical unit. The secondary chamber is vertical unit constructed as part of the stack. The primary chamber, which also serves as the "pyrolysis chamber", is operated under substoichiometric conditions at temperatures ranging from 650 to 870°C. Temperature control is provided by means of a burner and water sprays. Combustible gases generated in the primary chamber are burned in the secondary chamber, or "thermal reactor". The "thermal reactor" is operated at temperatures between 540 and 1090°C through the introduction of air and when needed, by supplying additional heat from an auxilliary burner.

# V. ENERGY CONSIDERATIONS

Refuse-to-energy projects are energy intensive. In fact, the question has arisen, and rightly so, as to the energy productivity of such systems. The final answer rests on the answer to the question: Does the amount of energy produced exceed that which is consumed in the transportation of the refuse and in the processing and conversion of it into energy? At this time, it can be stated categorically that the answer is in the affirmative. The correctness of the answer can be demonstrated by identifying, quantifying, and comparing the various energy consuming and energy producing activities of a typical energy conversion system.

## A. Net Energy Balance

Since the primary objective of a refuse-to-energy system is to utilize the heating value of refuse to produce a usable energy product, it is necessary to take into consideration the effectiveness of each system in terms of energy production. The effectiveness is a function of the net system output and is measured by the energy production ratio. The net energy is the initial value of the energy in the refuse minus energy requirements of the system and heat losses from it; the energy production ratio is the ratio of the value of the net output of energy from the system to the value of the energy content of input refuse. The higher the ratio, the more effective the system.

In order to develop the net system output for each system on a comparison basis, the following assumptions are made:

1. Net system output is calculated from the point at which the refuse enters the processing facility on through the production of the steam, and if RDF production is involved, through the point at which the RDF is converted to steam at the off-site conversion facility.
2. Incoming refuse has an energy value of 213 kJ/kg (4500 Btu/lb).
3. RDF is transported to the power plant to be combusted there.
4. The energy requirements of mobile equipment utilized at the proposed waste-to-energy conversion system must be included.

### *1. Net System Output*

The net system output can be calculated by deducting from the original energy value of the refuse (i.e., energy input) all process energy requirements and system losses. Process energy requirements include the energy in the form of the fossil fuel and electricity required to operate the energy conversion facility, and of the fossil fuel used in the transportation of residues at the disposal site. If RDF is used as the fuel, then the two preceding requirements must be added the energy used in the form of electricity in the operation of the refuse fuel processing plant, and in the form of the fossil fuel required to transport the RDF. System energy losses include the energy lost as heat during energy conversion, and if RDF is used, the form of the combustible material lost during the processing of the RDF. The net system energy output characteristic of each of the classes of energy production systems is summarized in Table 4. Data in the table were taken from References 4 through 10.

### *2. Energy Consuming Activities*

Energy consuming activities characteristic of each of the conversion systems are summarized as follows (listed in order of decreasing energy consumption):

**Table 4**
**NET SYSTEM ENERGY OUTPUT PER UNIT INPUT OF REFUSE[4-10]**

| | Waterall incineration of raw refuse | | Modular incineration | | RDF Typical | | RDF Enhanced[a] | |
|---|---|---|---|---|---|---|---|---|
| | (Btu/lb) | (J/g) | (Btu/lb) | (J/g) | (Btu/lb) | (J/g) | (Btu/lb) | (J/g) |
| Input | 4500 | 10,480 | 4500 | 10,480 | 4500 | 10,480 | 4500 | 10,480 |
| Front-end facility | | | | | | | | |
| Energy requirements | — | — | — | — | 70 | — | 70 | — |
| Losses | — | — | — | — | 586 | — | 1688 | — |
| Combustion facility | | | | | | | | |
| Energy requirements | 150 | — | 173 | — | 50 | — | 40 | — |
| Losses | 1850 | — | 1754 | — | 1214 | — | 591 | — |
| Output | 2500 | — | 2573 | — | 2580 | — | 2111 | — |

[a] May be prepared by pre- or posttrommeling. Transportation of solid fuel and/or residues is not included.

Modified from references 4 and 10.

1. Waterwall boiler power plants
   - Conversion (power and fuel demand)
   - Storage (power demand)
   - Conveyance (power demand)
   - Waste-oil processing (power demand)
   - Operation of rolling stock (fuel demand)
   - General purpose lighting, etc. (power demand)
2. RDF production facilities
   - Shredding (power demand)
   - Compaction—dRDF production is required for distant power plant (power demand)
   - Trommeling (power demand)
   - Conveyance (power demand)
   - Magnetic separation (power demand)
   - Operation of rolling stock (fuel demand)
   - General purpose lighting, etc. (power demand)
3. Modular incinerator/power plants
   - Conversion (power and fuel demand)
   - Operation of rolling stock (fuel demand)
   - Trommeling (power demand)
   - Conveyance (power demand)
   - Magnetic separation (power demand)
   - General purpose lighting, etc. (power demand)

*3. System Comparison*

The systems are ranked according to energy production effectiveness in Table 5. In the table is given the energy production ratio for each system.

## VI. MASS AND ENERGY BALANCE

The mass and energy balances to be expected with a refuse to energy plant may be typified by those for a 91-Mg/day (100 ton/day) plant. In developing the balances described in this section, data pertaining to refuse throughput and to operation were

**Table 5**
**ENERGY CONVERSION RATIOS OF COMMON SYSTEMS**

| System | Energy conversion ratio (%) |
|---|---|
| Waterwall combustion of unprocessed refuse | 56 |
| Modular incinerators | 57 |
| RDF production | |
| Typical | 57 |
| Enhanced | 47 |

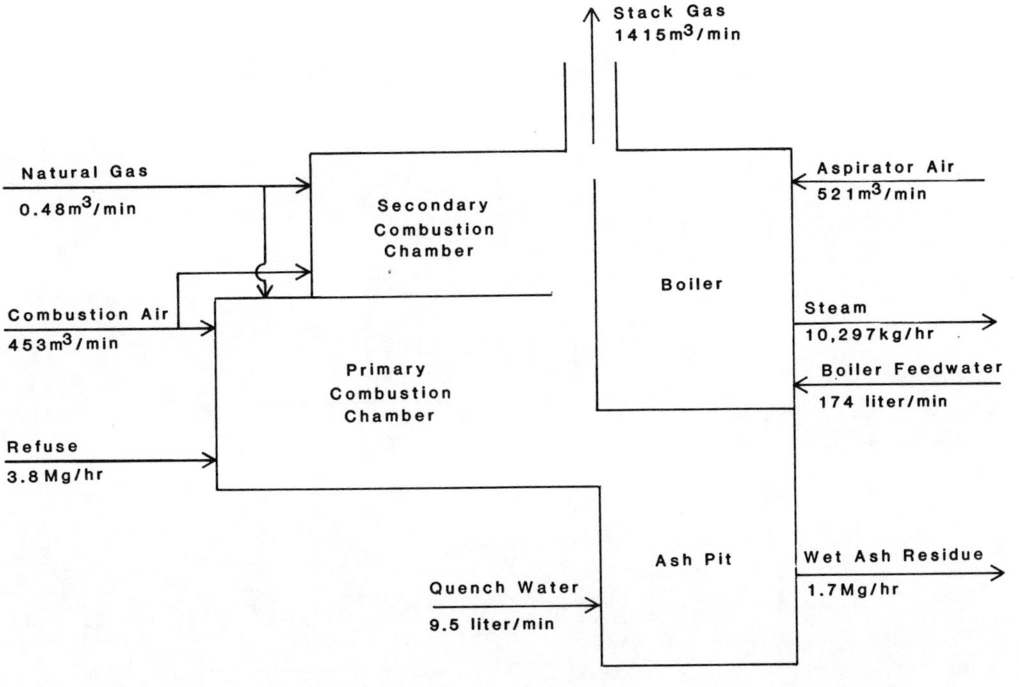


FIGURE 2. System diagram and mass flows for 91-Mg/day waste-to-energy system.

taken from References 4 and 5 and were modified and augmented as deemed appropriate on the basis of observational experience.

Inputs to the plant are as follows: (1) refuse at 3.8 Mg/hr (4.2 tons/hr); (2) combustion air supplied at standard temperature and pressure (STP) at 453 m$^3$/min (16,000 ft$^3$/min); (3) auxiliary fuel (natural gas) supplied at STP 0.48 m$^3$/min (17.0 ft$^3$/min); (4) aspirator air supplied at STP 521 m$^3$/min (18,400 ft$^3$/min) to the boiler section; (5) cooling water (15.6°C) for ash quenching supplied at 9.5 ℓ/min; and (6) condensate return plus makeup water (total boiler feed water) supplied at 174 ℓ/min. Temperature of the feedwater is 15.6°C. Mass outputs from the system are steam at 10,297 kg/hr (22,700 lb/hr), combustion gases (141°C) exiting to the atmosphere at 1415 m$^3$/min (50,000 ft$^3$/min), and a wet ash residue at 1.7 Mg/hr (1.9 tons/hr).

The principal components of the plant and the mass flows through it are presented in Figure 2; mass input and output values are listed in Table 6. Constituent dimensions of

**Table 6**
**MASS UNIT THROUGHPUTS FOR A 91-Mg/day (100 ton/day) WASTE-TO-ENERGY SYSTEM**

| | Throughput | | |
|---|---|---|---|
| | Mg/ Mg refuse | Ton/ton of refuse | Percent |
| Input | | | |
| Refuse | 1.0 | 1.0 | 4.5 |
| Combustion air | 9.5 | 8.6 | 38.5 |
| Natural gas | < 0.01 | < 0.05 | < 0.05 |
| Aspirator air | 10.8 | 9.8 | 43.9 |
| Residue cooling water | 0.2 | 0.2 | 0.9 |
| Boiler feedwater | 3.0 | 2.7 | 12.1 |
| Total | 24.5 | 22.3 | 100.0 |
| Output | | | |
| Steam | 3.0 | 2.7 | 12.1 |
| Stack gas | 21.0 | 19.1 | 85.7 |
| Residue | 0.5 | 0.5 | 2.2 |
| Total | 24.5 | 22.3 | 100.0 |

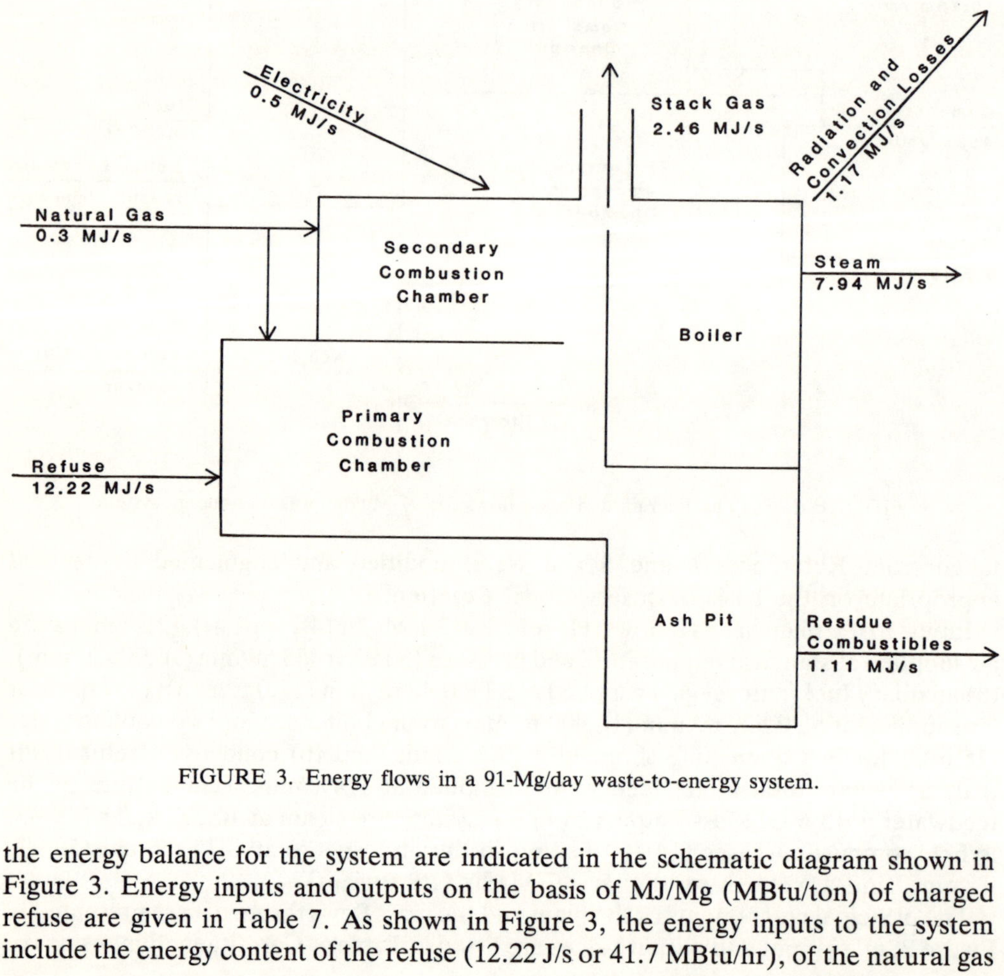


FIGURE 3. Energy flows in a 91-Mg/day waste-to-energy system.

the energy balance for the system are indicated in the schematic diagram shown in Figure 3. Energy inputs and outputs on the basis of MJ/Mg (MBtu/ton) of charged refuse are given in Table 7. As shown in Figure 3, the energy inputs to the system include the energy content of the refuse (12.22 J/s or 41.7 MBtu/hr), of the natural gas

**Table 7**
**ENERGY UNIT THROUGHPUTS FOR A 100-TPD WASTE-TO-ENERGY SYSTEM**

| | MJ/Mg of refuse | MBtu/ton of refuse | Percent |
|---|---|---|---|
| Input | | | |
| Refuse | 25.66 | 10.0 | 96.1 |
| Natural gas | 0.77 | 0.3 | 2.9 |
| Electricity | 0.26 | 0.1 | 1.0 |
| Total | 26.69 | 10.4 | 100.0 |
| Output | | | |
| Steam | 15.68 | 6.5 | 62.5 |
| Stack gas | 5.13 | 2.0 | 19.2 |
| Radiation and convection losses | 2.57 | 1.0 | 9.6 |
| Residue combustibles | 2.31 | 0.9 | 8.7 |
| Total | 26.69 | 10.4 | 100.0 |

(0.32 J/s or 1.1 MBtu/hr) supplied by auxiliary fuel burners, and of the electricity required to operate the system (0.15 J/s or 0.5 MBtu/hr). Auxiliary fuel requirements are determined by averaging the reported requirements of previously tested units.[5,11] Electric power consumption is based on operating data obtained from Reference 11. The energy content associated with the ash quench water and boiler feedwater, which enter the system at approximately 15.6°C, is not sufficiently great to warrant consideration.

Energy outputs from the system are (1) the 476 MJ/hr (27.1 MBtu/hr) of the 10,297 kg (22,700 lb) of steam (862 k/Nm², 125 psia) generated per hr; (2) the 147.6 MJ (8.4 MBtu) lost each hr in the form of the sensible heat of the stack gases; and (3) the 70.2 MJ (4.0 MBtu) lost from the system by way of radiation and convection. The sensible heat content of the gas is calculated on the basis of a stack gas temperature of 14.5°C. The heat content of the unburned combustible material remaining in the ash residue is derived from estimates of quantities and heating values of dry ash (see section entitled Residual Ash Considerations). Values given for radiative and convective heat losses are estimates based upon available information.

## VII. ECONOMIC FACTORS

A comparison between refuse-to-energy alternatives in terms of potential for cost effective operation requires the performance of a cost analysis. Capital costs, operating and maintenance costs, revenues generated from the sale of recovered products, and charges for the disposal of residues must be carefully evaluated.

### A. Capital and Operating Costs

Inasmuch as cost is perhaps one of the most important factors in decisions regarding the implementation of large-scale refuse-to-energy plants, it is very important to use proven methods for estimating and comparing costs. Unfortunately, a formidable obstacle militates against such a course of action. The obstacle is in the form of a general inability to make valid comparisons between cost estimates on the basis of numbers reported in the literature. The problem has a twofold origin: (1) the diversity of cost accounting methods employed; and (2) use of site-specific estimates. A practical result

**Table 8**
**TYPICAL STEAM PRODUCTION CAPACITY**

| System | Typical production (kg/kg MSW) | Theoretical[a] production (kg/kg MSW) |
|---|---|---|
| Waterwall combustion (with and without shredding) | 2.9 | 4.5 |
| Waterwall combustion (RDF) | 2.9 | 6.0 |
| Modular incinerator | 2.5 | 4.5 |

[a] H.V. (raw MSW) = 10,500 J/g;
H.V. (RDF) = 14,000 J/g;
efficiency = 100%.

of the diversity of methods is a difficulty in making a comparison between the estimates of two or more companies bidding on the same contract. A site-specific characteristic limits an estimate to the conditions peculiar to the situation for which the estimate was made. Since conditions vary significantly from situation to situation, the usefulness of the estimate is sharply reduced.

Capital costs of a 1100-Mg/day (1000 ton/day) waste-to-energy plant may range from 21 million to $55 million or more, depending upon the type of system chosen, construction, labor, materials, and equipment, and upon the costs of land site preparation. In terms of capital cost per megagram of daily capacity, the amounts range from $10,150 to $55,300/Mg daily capacity ($11,000 to $60,000/ton daily capacity). Many factors affect the annual operating costs of a waste-to-energy facility. Among them are the extent of the utilization of capacity, interest rate on borrowed funds, labor rates, and waste disposal costs. All of these factors are functions of plant reliability.

Finally, a few words of caution for one who is reviewing the cost analyses of any type of resource recovery facility are in order. Credits and revenues expected from the disposal of the wastes and the sale of recovered products (materials and energy) must be computed very carefully and conservatively. Unless serious negotiations have been carried out with the users of the facility as well as with the buyers of the recovered resources, values ranging from the best to the worst case should be used. Furthermore, another factor that affects the cost analysis is the quantity of waste to be processed by the facility. Typically, various systems are compared on the basis of dollars per megagram processed each day. Of course, any change in capacity, particularly a reduction in daily input, will increase the cost per unit weight processed.

### B. Markets for Recovered Energy

To realistically assess and make a meaningful comparison between the revenues projected for the various conversion systems, it is necessary to determine the extent of the potential demands of private firms and private and public institutions for the recovered energy. The need for the assessment and comparison arises from the fact that the size of the available market for the recovered energy and secondary materials produced by a project, and of the revenues therefrom, dictate the type and level of the processing required.

#### *1. Steam*

The amounts of steam that can be expected from the combustion of MSW or RDF are indicated by the data in Table 8. The 1980 market value of steam ranged from $0.007 to $0.010/kg ($0.003 to $0.0045/lb), or approximately $3 to $4.5/MBtu.

# VIII. POLLUTIONAL ASSESSMENT

## A. Introduction

Unless properly conducted, an incinerator operation can adversely affect air, water, and land resources. The adverse effect on the air would come from the indiscriminate release of stack exhausts containing gaseous and particulate pollutants generated in the combustion of refuse or other waste material. Water quality would be deteriorated through the discharge of quench water and water used in the boilers and waterwalls. The land resource could be deteriorated through the improper disposal of solid combustion residues. Of the three forms of pollution, air pollution was perhaps the most serious at the time of this writing. It most likely will continue to be a primary environmental deterrent to the implementation of an energy recovery system, although ash disposal was beginning to pose a real problem.

## B. Air Pollution

### *1. Introduction*

In this presentation, the exposition on air pollution and its control takes on the form of a case history based partially on an energy recovery project, which when implemented, would be sited in the San Diego (California) basin. While the system proposed for energy recovery in San Diego is based upon a 91-Mg/day (100 TPD) modular controlled-air incinerator designed for generating steam, the principles involved are more or less equally applicable to other forms of thermal energy recovery as well. An important step in assessing the environmental impact had or to be had by a plant with respect to air pollution is the making of a detailed evaluation of the air emissions coming from the plant in terms of their relation to state and local rules and regulations. The selection of a site in the San Diego Air Pollution Control District (SDCAPCD) is especially appropriate in this respect, because air pollution regulations in the district rank among the most severe in the nation.

### *2. Rules and Regulations*

Air pollution control rules and regulations are promulgated by way of federal, state, and county official publications (e.g., References 12, 13, 14). Generally, they are strongly time and place dependent. Obviously, a verbatim copying and a detailed analysis of the rules and regulations would not only be inappropriate, but also unnecessary. Hence, in this section, only a brief summary and analysis of rules and regulations that would apply to the proposed 91-Mg/day (100-ton/day) plant are given. In the presentation which follows, the time frame is 1979 to 1980, and the units of measurement are expressed according to the U.S. Customary System, i.e. the units used in drafting the regulations.

#### *a. Regulation IV*

**Rule 50-Visible Emissions**—No single source shall discharge into the atmosphere over an aggregate period of 3 min/hr or longer any air contaminant that registers darker than number 1 on the Ringleman scale.

**Rule 51-Nuisance**—Quantities of air contaminants that may cause injury, detriment, nuisance or annoyance to a considerable number of persons (or to the general public) shall not be discharged from a source. Nor shall air contaminants be emitted which endanger the comfort, repose, health, or safety of any such persons or cause injury or damage to business or property.

**Rule 53(a)-Sulfur Compounds**—No single source shall emit sulfur compounds calculated as $SO_2$, in concentrations in excess of 0.05% by dry volume.

**Rule 53(b)-Combustion Contaminants**—The maximum allowable concentration of combustion contaminants in the flue gas emitted by any single source is 0.1 gr/dscf corrected to 12% $CO_2$. The Environmental Protection Agency's (EPA) modified method-5 test procedure is used in determining concentrations, wherein the particulate catch from the probe, filters, and ice impinger are included. Both solid particles and condensibles are thus taken into account.

**Rule 54-Dust and Fumes**—No source shall discharge into the atmosphere in any one hour dust or fumes in excess of the amounts shown in Table 9. Process weight in the case of MSW incineration is the refuse throughput rate.

**Rule 58-Incinerator Burning**—Combustible refuse must be burned in a multiple-chamber incinerator or in equipment found in advance of use to be equally effective for the purpose of air pollution control by the Air Pollution Control Officer (APCO). A multiple-chamber incinerator has three or more refractory lined chambers interconnected by gas ducts.

**Rule 62(b)-Sulfur Content of Fuel**—Any liquid or solid fuel used in stationary fuel burning equipment shall contain no more than 0.5 percent sulfur by weight.

**Rule 68-Oxides of Nitrogen**—The maximum permissible concentration of $NO_x$ in the flue gas from a source with a maximum heat input rating of 50 million Btu/hr is 225 ppm (volume ) at 3% $O_2$ (430 g/m at 20°C).

### *b. Regulation X, Subpart E—Particulate Matter*

**Rule 260.50-Subpart E**—applicable to incinerators of more than 45 Mg/day (50 tons/day) charging rate.

**Rule 260.52**—The flue gas from an incinerator shall not contain particulate matter in excess of 0.08 gr/dscf corrected to 12% $CO_2$.

**Rule 260.53**—The owner or operator of an incinerator subject to the rules of Subpart E shall record the daily charging rates and hours of operation.

### *c. Regulation XI, Subpart C—National Emission Standard for Beryllium*

**Rule 361.32**—The maximum emission rate of beryllium from any source, including incinerators is 10 grams over a 24 hr period.

### *d. New Source Review Rules*

Air contaminants for which there is a state or federal ambient air quality standard are listed in Table 10.

**Rule 20.1**—This rule includes definitions and requirements regarding emissions calculations, emissions offsets and banking and exemptions.

Sections of Rule 20.1 that are applicable to the 91-Mg/day (100 TPD) incinerator/boiler system are as follows:

a. Best Available Control Technology (BACT)—maximum degree of air contaminant emission reduction which the air pollution control officer (APCO) determines is achievable on a case by case basis. In so doing, the officer is to take into account technology which although demonstrated, has not been necessarily proven in field application.
b. "Major Source"—the potential (uncontrolled) emission rate of any air pollutant for which there is a state or national ambient air quality standard is at least 91 Mg/yr (100 tons/ year), or the actual emission rate is at least 45 Mg/year (50 tons/year), 450 kg/d (1000 lb/day), or 45 kg/hr (100 lb/hr), whichever is greater. Quantities of air contaminants in the emissions from the 91 Mg/day (100 TPD) waste to energy plant would be well below those levels, provided, of course full recourse is made to

## Table 9
## MAXIMUM HOURLY PARTICLE EMISSION RATES[a]

| Process wt/hr | | Maximum wt discharge/hr | | Process wt/hr | | Maximum wt discharge/hr | |
|---|---|---|---|---|---|---|---|
| (kg) | (lb) | (kg) | (lb) | (Mg) | (lb) | (kg) | (lb) |
| 450 | 1000 | 1.26 | 2.80 | 4.5 | 10,000 | 4.50 | 10.00 |
| 900 | 2000 | 1.86 | 4.14 | 8.1 | 18,000 | 6.74 | 14.97 |
| 1800 | 4000 | 2.67 | 5.93 | 13.5 | 30,000 | 10.00 | 22.22 |
| 4050 | 9000 | 4.21 | 9.36 | 27.0 | 60,000 | 18.00 | 40.00 |

[a] Excerpted from Regulation IV, Rule 54.

## Table 10
## STATE OF CALIFORNIA AND FEDERAL AMBIENT AIR QUALITY STANDARDS FOR CERTAIN AIR CONTAMINANTS

| Pollutant | Averaging time | Calif. standards; concentration | Federal standards: Primary | Federal standards: Secondary |
|---|---|---|---|---|
| Photochemical oxidants (measured as ozone) | 1 hr | 0.10 ppm; (200 μg/m³) | 160 μg/m³; (0.08 ppm) | Same as primary standard |
| Carbon monoxide | 12 hr | 10 ppm; (11 mg/m³) | | Same as |
| | 8 hr | — | 10 mg/m³; (9 ppm) | primary |
| | 1 hr | 40 ppm; (46 mg/m³) | 40 mg/m³; (35 ppm) | standard |
| Nitrogen dioxide | Annual average | — | 100 μg/m³; (0.05 ppm) | Same as primary standard |
| | 1 hr | 0.25 ppm; (470 μg/m³) | — | — |
| Sulfur dioxide | Annual average | — | 80 μg/m³; (0.03 ppm) | — |
| | 24 hr | .05 ppm in comb w/ .10 ppm Ox or 100 μg/m³ TSP | 365 μg/m³; (0.14 ppm) | |
| | 3 hr | — | — | 1300 μg/m³; (0.5 ppm) |
| | 1 hr | 0.5 ppm; (1310 μg/m³) | — | — |
| Suspended particulate matter | Annual geometric mean | 60 μg/m³ | 75 μg/m³ | 60 μg/m³ |
| | 24 hr | 100 μg/m³ | 260 μg/m³ | 150 μg/m³ |
| Lead (particulate) | 30-day average | 1.5 μg/m³ | 1.5 μg/m³ | — |
| Hydrogen sulfide | 1 hr | 0.03 ppm; (42 μg/m³) | — | — |
| Hydrocarbons (corrected for methane) | 3 hr (6—9 a.m.) | — | 160 μg/m³; (0.24 ppm) | Same as primary standard |

**Table 10 (continued)**
**STATE OF CALIFORNIA AND FEDERAL AMBIENT AIR QUALITY STANDARDS FOR CERTAIN AIR CONTAMINANTS**

| Pollutant | Averaging time | Calif. standards; concentration | Federal standards | |
|---|---|---|---|---|
| | | | Primary | Secondary |
| Ethylene | 8 hr | 0.1 ppm | — | — |
| | 1 hr | 0.5 ppm | — | — |
| Visibility-reducing particles | 1 observation | Insufficient amount to reduce the prevailing visibility to 10 mi when the relative humidity is < 70% | | |

*Note:* Federal standard changed in 1979 to 0.12 ppm;
ppm = parts per million;
μ g/m³ = micrograms per cubic meter.

existing air pollution control technology. Therefore Rule 20.4, which has provisions regarding emissions offsets, lowest achievable emission rate and reasonable further progress, would not be cause for concern.

c. A "Stationary Source"—a unit or aggregation of units of nonvehicular air contaminant emitting equipment, which are all located on one property or adjoining properties under the same ownership and determined by the APCO to be related to one another through a similar product, raw material, or function. Obviously, the proposed 91-mg/day (100 ton/day) plant would fit under the definition.
d. Emission calculations—The maximum design capacity is used in determining emission from a source. Yearly emissions profiles will be used to determine the net increase in emissions from a new stationary source or modification.

**Rule 20.2-Best Available Control Technology (BACT)**—a. A new or modified source must be equipped with the best control technology available for any contaminant for which a state or national ambient air quality standard (criterion pollutant) has been established. The provision is appliable if the emissions are increased by the source in quantities exceeding 4.5 kg/hr (10 lb/hr) or 45 kg/day (100 lb/day), except in the case of CO, for which the limits are 45 kg/hr (100 lb/hr) or 450 kg/day (1000 lb/day). When interpreting BACT for a particular pollutant from a new source or modification, the air pollution control officer considers available control methods that are practical both economically and technically. b. If it can be shown that the cost of applying BACT in terms of dollars per unit of contaminant controlled is significantly greater than the cost of other control measures for the same contaminant, the air pollution control officer will not prescribe BACT for that contaminant. In lieu of BACT, emissions may be reduced in the same location as the new source (here, the 91–Mg/day plant) on a one-to-one basis. If it is determined that increased emissions from the new source shall not cause a local air quality problem, the in lieu reduction may be made in a different location in the air basin.

**Rule 20.3-Ambient Air Quality Analysis**—An authority to construct will be denied for any new or modified source that causes an increase in the emission of any criterion pollutant by more than 11 kg/hr (25 lb/hr) or 112 kg/day (250 lb/day) except CO for which the limit is 45 kg/hr or 450 kg/day), if the air pollution control officer determines that the emissions from the source will violate a state or national ambient air quality standard, or interfere with the attainment of such a standard for the contaminant in question. Computer modeling techniques, based on estimated emissions from the source will be used in the ambient air quality analysis.

Although the emissions of criteria contaminants from the new or modified source may not be expected to exceed the limits mentioned in the rule, the air pollution control officer can request an analysis of ambient air quality. Authority to construct will be denied if it is determined that a violation of a state or national ambient air quality standard will result.

The rule delineates the procedure for obtaining an authority to construct a new facility.

**Rule 20.4-Major Stationary Sources**—This rule would not be applicable to the proposed facility because the net increase in emissions due to the operation of the plant would not exceed the quantities named in the rule.

**Rule 20.6-Permit to Operate**—The process for obtaining a permit to operate is described in this rule.

**Rule 20.7-Prevention of Significant Deterioriation**—An authority to construct for a new or modified source which emits criteria pollutants will be denied by the APCO, unless the emissions neither directly nor indirectly cause a significant deterioration of existing air quality where the ambient air quality is better than an applicable air quality standard.

*3. Estimated Air Emissions*

Obviously, a MSW-fired boiler not equipped for air pollution control would emit more air pollutants than would an unequipped gas-fired unit. The difference between the two with respect to modular incinerators is exemplified by the data in Table 11. Emission data for units manufactured by Consumat® are available in two reports.[4,5] A report prepared for the Environmental Protection Agency,[4] includes data obtained from the monitoring of the air emissions from three modular Consumat® systems fired with municipal solid waste. The data are almost entirely limited to particulate emissions, although some information on visible emissions is included.

The most recently published air emission data are based on field tests conducted at a 45-Mg/day (50 ton/day) modular Consumat® incinerator boiler system located in North Little Rock, Arkansas.[5] The report contains emission data not only for particulate matter, but also for a number of other types of pollutants. Judging from the composition of the solid waste and the type of system employed at North Little Rock, the test data reported[6] would be indicative of the nature of the stack emissions from the proposed 91-Mg/day (100 ton/day) plant.

The controlled particulate emissions reported for the three modular Consumat® systems[4] were approximately the same as the maximum emission values named in Rule 53 of Regulation IV and in Subpart E of Regulation X. On the other hand, the uncontrolled particulate emissions reported for the modular incinerator boiler system located in North Little Rock[5] were well above the limits specified in the two rules. It should be pointed out that Subpart E of Regulation X would not be applicable to a 91-Mg/day (100 TPD) system composed of a group of incinerators, if each member of the group has a design capacity of less than 45 Mg/day (50 tons/day) and can be operated independently of the other members. (Consumat® and Comptro® recommend an arrangement of four 23-Mg/day (25 TPD) units for a 91-Mg/day (100 TPD) system.)

If the 1.49-kg/mg (3.03 lb/ton) uncontrolled particulate emission factor reported[5] is applied, the quantity of uncontrolled emission of particulates from the 91-mg/day (100 ton/day) plant would be in excess of the 3.97 kg/hr (8.83 lb/hr) limit set by Rule 54 of Regulation IV (see Table 9). Because the net increase in the uncontrolled particulate discharge rate due to the operation of the proposed plan would exceed the 113 kg/day

**Table 11**
**UNCONTROLLED EMISSION FROM A MSW-FIRED BOILER AND A GAS-FIRED BOILER**

| Pollutant species | MSW-fired boiler[a] | | Natural gas-fired boiler[b] | | Net increase | |
|---|---|---|---|---|---|---|
| | (kg/day) | (lb/day) | (kg/day) | (lb/day) | (kg/day) | (lb/day) |
| Particlate matter | 136 | 303 | 5 | 10 | 131 | 293 |
| $NO_x$ | 166 | 368 | 51 | 114 | 115 | 254 |
| $SO_x$ | 35 | 78 | < 1 | 1 | 34 | 77 |
| HC | 25 | 55 | 4 | 8 | 21 | 47 |
| CO | 45 | 100 | 9 | 19 | 36 | 81 |

[a] 91 Mg/day (100 ton/day).
[b] Rated at 12.22. MW/hr (41.7 MBtu/hr).[12]

(250 lb/day) limit of Rule 20.3 of the New Source Review, an ambient air quality analysis would have to be made before the authority to construct the proposed plant could be obtained. Furthermore, since the net increase would be greater than 45 kg/day (100 lb/day), the best available control technology for particulate removal would be required. Because of this requirement, manufacturers of both small- and large-scale refuse burning systems recommend the incorporation of particulate control equipment into their systems.

As reported,[5] uncontrolled $NO_x$ emissions exceed the limit of 225 ppm at 3% $O_2$ specified in Rule 58 of Regulation IV. But then, Rule 58 is applicable only to combustion sources that have a maximum energy input rating of 14.7 MW/hr (50 mbtu/hr) or more, whereas a 91-mg/day (100 TPD) system firing MSW (11,630 J/g, 5000 btu/lb) and operating 24 hr/day would have a heat input rating of less than 12.3 mol wt. (42 Mbtu)/hr. Therefore, the proposed plant apparently would not be within the scope of Rule 58.

However, on the basis of data reported,[5] the net increase in uncontrolled $NO_x$ emissions due to the proposed 91-mg/day (100 ton/day) plant would be in excess of the 113-kg/day (250 lb/day) limit established by Rule 20.3 of the New Source Review.

While the data presented[5] give reason for concluding that $NO_x$ control would be required for the proposed plant, it is not at all certain that the high emission factors for uncontrolled $NO_x$ emissions reported[5] are representative of all modular-controlled air systems. For instance, the high $NO_x$ emission factors reported for the North Little Rock facility may have been due to the poor operational (e.g. combustion) controls that were in use at the time of the testing. Manufacturers of small-scale waste-to-energy systems have always claimed that $NO_x$ emissions from their systems would be minor in extent. Unforunately, there always is the possibility that the local district may choose to regard data[5] as constituting the best available information for evaluation purposes until the test data become available.

Judging from available information, uncontrolled visible emissions from a 91-Mg/day (100 TPD) plant could possibly pose a problem. Although the standard set forth in Rule 50 of Regulation IV was not exceeded in the tests conducted on the three modular Consumat® incinerators,[4] other information[12,14] suggest that visiblity of emissions from waste-to-energy systems could present difficulties. Possibly, the problem could be alleviated through the use of a particulate control device. More intensive research on visible emissions from waste-to-energy systems and their control are needed before a ready solution is found.

The uncontrolled emission of lead from a 91-mg/day (100 ton/day) waste-to-energy plant could become a problem, even though the uncontrolled emission factor for lead reported[5] is indicative of an emission rate for a 91-mg/day (100 TPD) waste-to-energy system that would be well below the applicable standard of 3.97 kg/hr (8.83 lb/hr). (Lead emissions are restricted by the provisions of New Source Review Rule 20.2 and Rule 54 of Regulation IV.) The fact remains, however, that high concentrations of lead have been observed in the flue gases from refuse combustion operations.[15] The problem can be solved through the installation of a particulate control device, since lead is emitted from refuse-burning operations primarily as a particulate.

If emission data from the North Little Rock facility[5] can be taken as being representative of modular waste-to-energy systems, uncontrolled $SO_x$, CO, HC, and beryllium emissions should not exceed the regulatory limits. The net increase in daily discharges of $SO_x$, CO, and HC should be well below the 45-kg (100 lb) limit (450 kg, 1000 lb for CO) established in Rule 20.2 of the New Source Review and the additional restrictions set up for $SO_x$ by Rules 53(a) and 62 of Regulation IV. According to Rule 53(a), the concentration of $SO_x$ in the flue gas must be less than 500 ppm. Rule 62 of Regulation IV specifies that solid fuels may contain no more than 0.5% sulfur by weight. Inasmuch as the sulfur content of municipal refuse is very low and the auxiliary fuel requirement for modular controlled-air incinerators is minimal, the standards of Rules 53(a) and 62 would easily be met by the 91-mg/day (100 TPD plant). Emissions of hydrogen sulfide are limited to a net release of 45 kg/day (100 lb/day) by Rule 20.2 of the New Source Review. No $H_2S$ was detected during the tests run at the North Little Rock facility.[5] Hydrogen sulfide emissions from the proposed plant would not be significant in amount.

Beryllium emissions are restricted to 10 g/day (Regulation X, Subpart C, Rule 361.32). Data from the North Little Rock facility[5] indicate that the emission rate of beryllium from a 91-mg/day (100 TPD) plant would be less than 2% of the limit.

Inasmuch as the state of California has established ambient air quality standards with respect to ethylene and visibility reducing particles, the limits set forth in Rules 20.2 and 20.3 of the New Source Review also apply to the two contaminants. Presently, no emission data from modular waste-to-energy systems are available that deal with these two contaminants.

*4. Environmental Impact Report/Study and Permit Requirements*

Although an air pollution control district may not be designated as the lead agency for a county in the development of an environmental impact report/study (EIR/EIS), a district nevertheless may insist upon a report in its review of a proposed waste-to-energy plant. Generally, it can be assumed that an EIR/EIS would be required for a major combustion source. With respect to the San Diego District, the requirement of obtaining permits for the construction and operation of equipment that emits or controls air contaminants is set forth in Rule 10 of the district's rules and regulations. Individual operating permits would be required by the district for each proposed incinerator that could be operated independently. The requirements for filing for operational permits are specified in Rule 14. Information must be included in the permit application on the use of equipment to comply with the applicable district rules and regulations and of the extent of the compliance made possible by the use of that equipment. In Rule 19 is specified the extent and nature of the provision and maintenance required in terms of source sampling and testing facilities for equipment for which a permit is required, if such a requirement is made a part of the authority to construct or of a permit to operate. The procedures for obtaining an authority to construct and to operate are included in Rules 20.3 and 20.6, respectively, of the New Source Review.

Finally, according to Rule 58 of Regulation IV, if a controlled-air incinerator were chosen for the 91-Mg/day (100 TPD) plant, it would have to be demonstrated to the air pollution control officer as being as effective as a multiple chamber incinerator in terms of air pollution control. On the basis of the emission factors reported,[5,16] it may be concluded that compliance with the rule would not pose a problem.

The nuisance problems named in Rule 51 of Regulation IV will not occur if the proposed waste-to-energy system is properly designed, enclosed in a building, and operated correctly.

*5. Control Equipment and Cost*

Upon the basis of the projected amounts of uncontrolled emissions and the applicable rules and regulations established by the San Diego County Air Pollution Control District, the use of one and possibly of two emission control methods would be required in the proposed 91-Mg/day (100 TPD) plant to satisfy the emission standards. Equipment for particulate control would most certainly be required. Hence, most manufacturers of controlled-air incinerators currently recommend the installation of a particulate control device such as a baghouse or an electrostatic precipitator (ESP). If the need is one of controlling the emission of $NO_x$, it will be hard to meet. The most that can be said of existing $NO_x$ control methods for controlled air incineration in particular, and for combustion systems in general, is that the methods are ill-defined, potentially expensive, and are the subjects of a great deal of debate.

***a. Particulate Control***

The average particle size of the particulates discharged in the stack exhaust from a waste-fired modular incinerator is on the order of 0.3 μm.[5] Judging from reported data on uncontrolled emissions,[5,11] a particulate removal efficiency of 80% would be needed to meet the San Diego particulate standards. Such an efficiency could be attained with the use of a baghouse filter. Baghouses are among the most efficient particle collection devices presently available, and effectively control particles that are sized in the range of 0.2 to 2 μm.

Problems may be encountered when a baghouse is used in conjunction with a controlled-air incinerator. The main sources of the problems are the high flue-gas temperature and the presence of certain chemicals. Flue-gas temperatures in waste-fired modular incineration/boiler units are approximately 150°C.[4,5] "Nomex"® and fiberglass are two fabric filtering materials that can withstand a temperature of 150°C. The maximum temperature for fiberglass is as high as 260°C. However, "Nomex"® is much more resistant than is fiberglass to the chloride and flouride gases present in combustion products formed in the burning of solid waste. Fiberglass bags can be severely attacked by chloride (HCl) and flouride (HF) gases. The cost of frequent bag replacement in terms of direct labor cost and equipment downtime may render a baghouse equipped to use fiberglass bags economically impractical. The use of "Nomex"® is a gamble, inasmuch as the temperature limit (190°C) of the material is only slightly higher than that of the temperature of a typical flue gas.

In 1980, capital costs for baghouses were about $2120 per actual cubic meter per second ($10 per actual cubic feet per min). Assuming a gas flow rate of 23.5m³/s (50,000 acfm) and a refuse charging rate of 91-Mg/day (100 TPD), the installed capital cost for a baghouse for a 91-Mg/day (100 TPD) incinerator was approximately $500,000. The annual operating and maintenance costs were on the order of $50,000. As far as operating and maintenance costs of the baghouse are concerned, they would be functions of the temperature of the flue gas and the concentration of corrosive gases in the flue gas.

The operating and maintenance costs for an electrostatic precipitator (ESP) would be lower than those projected for a baghouse. The installed capital cost of an ESP for the 91-Mg/day (100 TPD) plant in 1979 to 1980 was approximately $350,000 to $450,000. The annual operating and maintenance cost for an ESP sized for a 91-Mg/day (100 TPD) plant was in the range of $20,000 to $40,000. While on the subject of electrostatic precipitators, it should be mentioned that in some large waste-to-energy installations (i.e., processing greater than 181 Mg/day), the performance of the precipitators has been less than satisfactory. One specific problem area was an incompatibility between the electrical characteristics of refuse fly ash (primarily its resistivity) and the operation of the precipitator.

***b. $NO_x$ Control***

Technology for the control of $NO_x$ emissions is for the most part concerned with utility boilers, process heaters, and internal combustion engines. Operational control methods include flue gas recirculation, burner modifications, and reduced air preheat. Catalytic denitrification by way of injection of ammonia has been tried in certain boiler facilities. The use of catalytic methods for control is accompanied by an element of uncertainty that arises from the possibility of the catalyst being "poisoned" by certain constituents of the flue gas. Costs (1980) for operational modifications of a controlled-air incinerator probably would be on the order of $165/Mg of $NO_x$ removed.

The economic exemption from the best available control technology requirement made in paragraph b of Rule 20.2 of the New Source Review may be applicable to the necessity of providing means of $NO_x$ control. For example, to qualify for the exemption in certain air pollution control districts, it would be necessary to convince the air pollution control officer that the costs incurred in the application of best available technology to the removal of $NO_x$ from the stack discharge of the proposed incinerator (boiler) would be much higher than would be the case with other sources of the pollutant (e.g., utility boilers, automobiles). Not to be overlooked is the "in lieu" exemption named in paragraph b of Rule 20.2. The exemption is in the form of $NO_x$ offsets that can be attained on a one-to-one basis, i.e., one unit of $NO_x$ removed elsewhere for one emitted by the proposed waste-to-energy system.

*6. Summary*

From the preceding presentation, it is evident that to meet the demands of rules and regulations such as those laid down by the San Diego Air Pollution Control District, the emission of particulates from a waste-to-energy plant would have to be controlled. Of the control devices available or potentially so at this time, the electrostatic precipitator and the baghouse would seem to be the best choices for particulate control.

In the discussion of system economics, a capital cost of $450,000 and a range of $20,000 to $50,000 for annual operating and maintenance costs have been assumed. These costs can be considered as being representative of both the electrostatic precipitator and the baghouse particulate control systems.

Due to a paucity of data on $NO_x$ control for waste-fired systems and to the fact that an exemption could possibly be attained on the basis of extremely high costs, no economic analysis of $NO_x$ control is developed herein.

**C. Residual Ash Considerations**

*1. Characteristics*

All modular controlled-air incinerators currently available are equipped for ash removal. Most large units have an automatic ash removal system and an ash quenching pit. Generally speaking, four types of solid residues are characteristic of the incineration process, namely, unburned waste, inert materials, ash, and fly ash. The four residues collectively are known as "ash". In weight they are equal to approximately

**Table 12**
**CHARACTERISTICS OF INCINERATOR ASH[a]**

| | | |
|---|---|---|
| Weight reduction[b] | | |
| Dry ash (%) | | 70 |
| Wet ash (%) | | 55 |
| Volume reduction, % on wet basis[c] | | 90—95 |
| Wet ash density, kg/m$^3$, (lb/yd$^3$) | 890 | (1500) |
| Heating value of dry ash, J/g, (Btu/lb) | 2328—4656 | (1000—2000) |
| Nonburned combustible content of ash (% dry weight) | | 5—15 |

[a] Based upon in-house information and data from References 4 and 5.
[b] As indicated by weight of ash formed.
[c] Based on volumes of input wastes and that of the ash.

**Table 13**
**CHARACTERISTICS OF INCINERATOR WASTEWATER FROM SIX PLANTS[a]**

| Characteristic | Quench water (range) | Fly ash effluent (range) |
|---|---|---|
| pH | 10.4—11.0 | 4.6—4.8 |
| Dissolved solids, mg/ℓ | 2551—4283 | 5882—8838 |
| Suspended solids, mg/ℓ | 372—879 | 325—1351 |
| Total solids (% volatile) | 27—36 | 24 |
| Hardness ($CaCO_3$) mg/ℓ | 739—904 | 2632—3437 |
| Sulfate ($SO_4$) mg/ℓ | 242—371 | 725—1250 |
| Phosphate ($PO_4$) mg/ℓ | 23.3—33.8 | 13—51 |
| Chloride (Cl) mg/ℓ | 393—868 | 2297—3543 |
| Alkalinity ($CaCO_3$) mg/ℓ | 465—671 | 1.33—23.0 |

[a] Plant capacities were from 100 Mg (110 tons) to 450 Mg/day (500 TPD).

Modified from DeMarco, J., Keller, D. J., Leckman, J., and Newton J. L., Incinerator Guidelines — 1969, U.S. Department of Health, Education and Welfare, Bureau of Solid Waste Management, 1969; reprinted by the U.S. E.P.A., 1973.

20 to 30% of the incoming waste. Since unburned wastes in the ash are those which escaped being burned during the course of the incineration process, obviously the amount of ash in that form decreases with increase in the efficiency of combustion.

An indication of pertinent characteristics of the ash is given by the data in Table 12. The bulk density of wet ash is approximately 890 kg/m$^3$ (1500 lb/yd$^3$). The ash may contain from 5 to 15% nonburned combustible matter, and as such may have a residual heating value of 2328 to 4656 J/g (1000 to 2000 Btu/lb).

Incinerator ash may contain significant concentrations of heavy metals. The only available data for a controlled air incineration system[5] indicate that concentrations of Cd, Co, Pb, and Zn may exceed 1000 ppm.

*2. Disposal*

On the basis of the data in Table 13, probably as much as 41 Mg (45 tons) (wet weight) of ash will have to be removed daily from a 91-Mg/day (100 TPD) waste-to-energy facility. (The volume will be 46 m$^3$ or 60 yd$^3$.) The ash would have to be disposed of at a

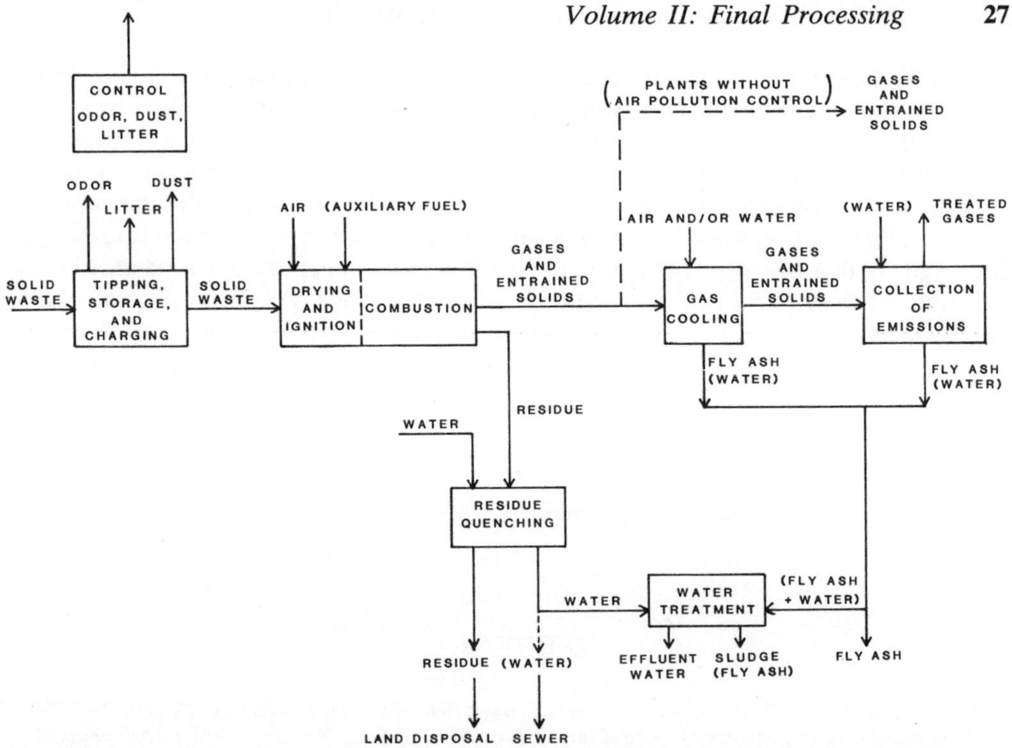


FIGURE 4. Schematic diagram of processes, inputs, and outputs of a typical incineration plant.

Class I disposal site, because the California State Water Resources Control Board currently classifies community incinerator ash as a Group 1 waste as specified in Title 23, Chapter 3 of the California Administrative Code. (A Group 1 waste is one that contains or consists of toxic substances.)

Fees for ash disposal most probably would include two additive charges. The first charge, or nominal fee, would be based upon quantity (e.g., weight) of material requiring disposal. The second charge, or special handling fee, would be based upon the judgment of the landfill operator as to the nuisance potential of the material. Nuisance potential, in the case of incinerator ash, includes the propensity of the material to being blown around the landfill, the prevention of which may involve some form of special handling. The special handling charge would commensurate with the degree of handling required. At this time, the extent of such special handling for ash from a 91-Mg/day (100 ton/day) plant is a matter of conjecture.

Ash disposal charges quoted in the San Diego area in 1980 were $11.55/Mg ($10.50/ton) of material disposed. The special handling charge was as high as $500 per load. The latter charge on a wet weight basis is $12.22/Mg ($11.11/ton) of ash disposed, assuming that one load is the total ash disposed of at the landfill on a given day. The total fee for residual ash disposal at a Class I landfill was the equivalent to $5.20 to $10.69/Mg ($4.73 to $9.72/ton) of refuse burned.

If incinerator ash is classified as being a nonhazardous Group 2 waste, landfill costs would be reduced significantly.

### D. Wastewater Considerations

Wastewater resulting from the operation of a waste-to-energy system consists of ash quench water, water used in air pollution control, and water used in plant maintenance and sanitation. The uses of water in the incineration process, and hence the manner in which it is polluted. (Figure 4) The capacities of the plants were from 100 to 450

Mg/day.[17] The extent of the contamination of the water is indicated by the data in Table 13. From the data it becomes readily apparent that governmental regulations (local, state, and federal) would bar the discharge of untreated incineration wastewater into the environment. In a practical operation, the wastewater could be subjected to simple treatment (e.g., passed through a settling tank) and then reused in the incineration process. The recycling would result in a substantial reduction in the amount requiring discharge. With a 100-Mg/day facility, the discharge of quench water would only be on the order of 946 to 3785 ℓ/day, and of equipment and tipping floor washwater, only 1135 to 3785 ℓ/day.

## REFERENCES

1. **Goodrich, W. F.,** *The Economic Disposal of Towns Refuse,* P. S. King and Son, London, 1901, 340.
2. **Weinstein, N. J.,** Municipal-Scale Thermal Processing of Solid Wastes, U. S. EPA Report No. EPA/530/SW-1330, 1977, 349.
3. **Bozeka, C. G.,** *Nashville Incinerator Performance Tests,* Babcock and Wilcox, North Canton, Ohio, 1975.
4. Evaluation of small modular incinerators in municipal plants, Ross Hofmann Associates, Coral Gables, Fla., prepared for EPA Office of Solid Waste Management, Contract 68-01-3171, 1976.
5. Small modular incinerator systems with heat recovery, Systems Technology Corp., Xenia, Ohio, prepared for EPA Office of Water and Waste Management, Contract 68-01-3889, 1979.
6. **Savage, G. M., Diaz, L. F., and Trezek, G. J.,** RDF: quality must precede quantity, *Waste Age,* 9(4), 100, 1978.
7. **Levy, S. J. and Rigo, H. G.,** Resource Recovery Plant Implementation: Guides for Municipal Officials—Technologies, U. S. EPA, Report No. SW-157.2, 1976, 81.
8. **Trezek, G. J., Diaz, L. F. and Savage, G. M.,** Prediction of the impact of screening on refuse-derived fuel quality, prepared for the Electric Power Research Institute, Report No. EPRI FP-1249, November 1979, 68.
9. St. Louis Demonstration Final Report: Refuse Processing Plant Equipment, Facilities and Environmental Evaluations, EPA-600/2-77-155b, December 1977.
10. **Even, J. C., Adams, S. K., Gheresus, P., Joensen, A. W., Hall, J. L. Fiscus, D. E. and Romine, C. A.,** Evaluation of the Ames Solid Waste Recovery System: Part I—Summary of Environmental Emissions, Equipment, Facilities, and Economic Evaluations, November 1977.
11. Status of Technology in the Recovery of Resources from Solid Wastes, County Sanitation District, Los Angeles County, Calif., January 1976.
12. Air Pollution Control Regulations, State of California Air Resources Board, Sacramento, Calif., 1979.
13. Air Pollution Control Regulations, County of San Diego Air Pollution Control District, San Diego, Calif., January 1979.
14. **Hammond, W. F.,** On-Site Evaluations of Refuse-Energy Systems, Los Angeles County Air Pollution Control District, May 1975.
15. **Olexsey, R. A.,** Environmental impact of conversion of refuse to energy, paper presented at the International Recycling Congress, Berlin, 1979.
16. Compilation of Air Pollutant Emission Factors, EPA Publ. AP-42, Office of Air and Waste Management, Research Triangle Park, N.C., May 1978.
17. **DeMarco, J., Keller, D. J., Leckman, J., and Newton, J. L.,** Incinerator Guidelines—1969, U. S. Department of Health, Education and Welfare, Bureau of Solid Waste Management, 1969; reprinted by the U. S. E.P.A., 1973.

Chapter 2

# PRODUCTION AND UTILIZATION OF REFUSE-DERIVED FUEL

## I. INTRODUCTION

Because the technologies for producing and using waste fuels, particularly refuse-derived fuel (RDF), are quite new, experience with full-scale firing of RDF has been very limited, and reported data are presently few and far between. In the U.S. the first full-scale utilization of RDF took place in a project begun in 1972 and conducted jointly by the city of St. Louis and the Union Electric Company.[1] In the project, RDF and powdered coal were fired in one of the utility boilers. Since that first venture, a number of cities have adopted or are endeavoring to adopt systems similar to the one used in the St. Louis project. Projects, existing or planned, are listed in Table 1. Four of them are briefly described at the end of this chapter.

## II. PREPARATION OF RDF

In the early stages of the operation of the St. Louis project, the only treatments accorded the raw refuse (MSW) in the production of RDF were size reduction and magnetic separation. A serious disadvantage of such a simplified approach is that noncombustible (e.g., glass, nonferrous metals, grit) and difficultly managed materials (e.g., textiles) become a part of the RDF, and thereby lead to stoppages and excessive wear and tear of the pneumatic and feeding systems to the boiler. In the St. Louis project, an air classification step was introduced before the latter problems had time to develop. Thus, in the final version of the St. Louis system, RDF production involved three basic unit processes: size reduction, magnetic separation, and air classification.[1]

In undertakings that came into being in other parts of the U.S. after the termination of the St. Louis project, other unit processes were joined to the three basic unit processes with the hope of further beneficiating the RDF product, and of reclaiming additional useful materials from the refuse.[3,4] Of the new unit processes, trommeling and screening proved to be more successful.[5] (Although "trommeling" is a form of screening, and thus the expression "trommeling and screening" is redundant, the expression, nevertheless, is used in this book to give emphasis to trommeling.) Consequently, most refuse processing plants designed to produce RDF now include trommeling in addition to the usual size reduction, magnetic separation, air classification, and screening. The complete processing operation involves a minimum of one pass through each of the five unit processes.

### A. Types of RDF

Types of RDF produced in modern plants are functions of equipment design, arrangement (e.g., sequence), and operation. They generally fall into three broad groups: coarse RDF, "fluff RDF", and densified RDF ("d-RDF"). The unit processes typically involved and their sequence in the production of each of the three types are shown in Figure 1. The primary size reduction prescribed for all three types is a reduction to a particle size of less than 10 to 15 cm.[6]

According to Figure 1, the final step in the production of coarse RDF is the removal of the "lights" produced in air classification. Thus, the RDF in this case is the light fraction of the refuse. (In the remainder of this section, "coarse RDF", "RDF", "lights", and "light fraction" are used interchangeably.) The position of the magnetic

**Table 1**
**EXISTING AND PLANNED SOLID FUEL PROJECTS AS OF 1981**

| Location | Fuel type | Capacity (Mg/d) | Capital cost ($/input Mg) | Status |
|---|---|---|---|---|
| Bridgeport, Conn. | Eco-Fuel II | 1630 | 58.3 | CC |
| Wilmington, Del. | RDF | 910 | 78.4 | CB |
| Lakeland, Fla. | RDF | 270 | 5.5 | CB |
| Ames, Iowa | RDF | 180 | 6.8 | O |
| New Orleans, La. | RDF | 635 | 10.0 | O |
| Chicago, Ill. | RDF | 910 | 20.9 | OS |
| Baltimore Co, Md. | RDF | 545—1360 | 9.2 | O |
| East Bridgewater, Mass. | Eco-Fuel | 500 | 11—13 | O |
| Newark, N.J. | Eco-Fuel | 2720 | 77.0 | SC |
| Monroe, Co., N.Y | RDF | 1815 | 66.2 | O |
| Akron, Ohio | RDF | 910 | 60.5 | O |
| Lane Co., Ore. | RDF | 450 | 2.3 | CC |
| Tacoma, Wash. | RDF | 450 | 2.8 | O |
| Madison, Wis. | RDF | 360 | 2.8 | O |
| Milwaukee, Wis. | RDF | 1450 | 19.8 | O |

*Note:* Key: CC = construction completed;
CB = construction begun;
O = operational;
L = fuel not sold;
OS = off-stream;
SC = site preparation completed.

Modifed from *N.C.R.R. Bull.* 10(3), 1980.

separation step in the sequence is flexible in that it may either precede or succeed air classification. In use, coarse RDF is injected into the furnace and fired in "semisuspension".[7]

"Fluff RDF", as shown in Figure 1, is produced through the addition of screening and secondary size reduction to the four basic unit processes applied in the production of coarse RDF. The final particle size of "fluff RDF" usually is 2.0 to 2.5 cm. Characteristics of "fluff RDF" are indicated by the two sets of data in Table 2.

A variation of "fluff RDF" bears the designation "dust RDF".[8] In the production of "dust RDF", the secondary size reduction is accomplished by means of a ball mill. In the operation, the abrading elements are heated and an embrittling agent is introduced. The milled product is screened and cooled. Characteristics of the product are indicated by the data in Table 3.

### B. Densified RDF

Densified RDF ("d-RDF") is the term applied to "fluff RDF" that has been extruded into pellets or cubettes (cf. Figure 1). The growing recognition of the advantages in terms of ease and cost of transportation and storage, as well as of adaptability to certain types of firing that attend the use of densified RDF, has brought with it an intensification of interest in the process.[4,9] A result of the relative recency of this recognition is the fact that as of this writing, published data on the preparation of densified RDF and its characteristics are mostly to be found in a report prepared by Trezek et al.[10] Accordingly, data and information presented in this section are based upon the results reported by Trezek et al. in the report.

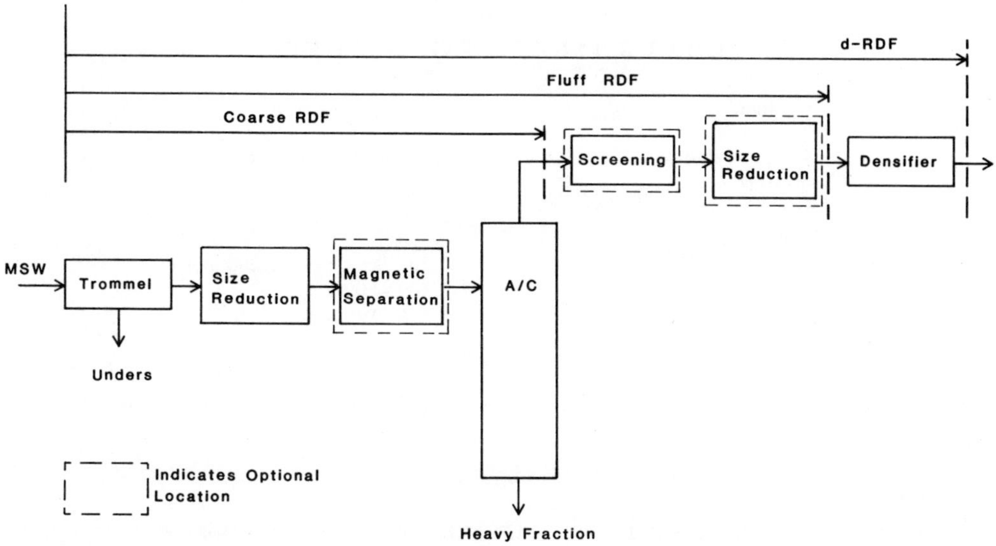


FIGURE 1. Typical sequence of unit processes for RDF production.

**Table 2**
**CHARACTERISTICS OF "FLUFF RDF" PRODUCED AT AMES AND ST. LOUIS**

| Item, analysis of RDF | Mean values | | Difference |
|---|---|---|---|
| | Ames | St. Louis | (Ames vs. St. Louis) |
| Bulk density (kg/m³) | 128.3 | 130.1 | −1.4 |
| Heating value (kJ/kg) | 13,050 | 10,636 | +22.7 |
| Moisture (%) | 23.03 | 26.60 | −13.4 |
| Ash (%) | 17.37 | 21.70 | −20.0 |
| Volatile matter (%) | 54.65 | 43.60 | +25.3 |
| Fixed carbon (%) | 4.94 | 8.10 | −39.0 |
| Carbon (%) | 30.86 | 26.00 | +18.7 |
| Hydrogen (%) | 4.77 | 3.79 | +25.9 |
| Oxygen (%) | 22.88 | 21.20 | +7.9 |
| Sulfur (%) | 0.43 | 0.18 | +138.9 |
| Chlorine (%) | 0.24 | [a] | [a] |
| Nitrogen (%) | 0.42 | 0.53 | −20.8 |
| Particle size (geometric mean diameter [mm]) | 10.7 | 7.4 | +44.6 |

[a] Chlorine not determined during St. Louis tests.

Modified from St. Louis Demonstration Final Report: Refuse Processing Plant Equipment, Facilities and Environmental Evaluations, EPA-600/2-77-155a, December 1977; Even, J. C. et al., Ames Evaluation of the Ames Solid Waste Recovery System, EPA Grant No. R803903010, December.

In their study, Trezek et al.[10] were concerned with the effects of moisture content and throughput rate on the specific energy requirements of the densification process as applied to RDF. Both screened and unscreened air-classified "lights" were tested. Both types of lights were produced through conventional front-end processing. Throughput from the pellet mill was measured by collecting all pellets extruded during 1- and 2-min

**Table 3**
**CHARACTERISTICS OF DUST RDF**

| Item | Measure |
|---|---|
| Heating value | 17.4—18.62 kJ/g (dry) |
| Moisture content | <5% |
| Ash content | <15% |
| Sulfur content | <1% |
| Particle size | 50% smaller than 200 mesh |
| Density | 400—800 g/ℓ |
| Energy required to produce | 516 kWh/Mg |

intervals. Fuel characteristics of two of typical samples of densified RDF are listed in Table 4.

*1. Throughput*

Results obtained in the testing indicate that the specific energy apparently decreases with increase in throughput rate, as is shown by the curves in Figure 2. According to the figure, the specific energy increased sharply as the throughput approached 100 kg/hr (0.1 tons/hr), and became much greater at lesser throughputs. The trend was found to have prevailed throughout the range of pellet densities tried in the study, namely, from 993 kg/m$^3$ to 1346 kg/m$^3$, inclusively. The data can be expressed as a power law by fitting a relationship of the type $Y = AX^b$, wherc the coefficients A and B vary slightly with density.

*2. Moisture*

Within the range of pellet moisture contents tested in the study, i.e., 13.6 to 26.4% inclusively, the energy requirement declined with increase in moisture, probably because of the "lubricating" effect of the water, as well as a lessening of the resilience of the paper fibers. The fact that the trends for moisture content and throughput rates were parametric is shown by the relative positions of the three curves in Figure 2. The effect of increasing moisture content upon the density of the pellets was in the opposite direction. At a moisture content of 14%, the density was on the order of 1300 kg/m$^3$ (81 lb/ft$^3$), whereas at 26% moisture, it dropped to 1100 kg/m$^3$ (69 lb/ft$^3$). The relation is shown by the curve in Figure 3.

According to the data in Table 5, in which are listed the energy requirements of the major unit processes involved in the production of densified RDF,[4] the energy consumed in densifying RDF was slightly more than 1.5 times that used in the air classification step. Data in the table also indicate that the energy consumed in densification is on the order of 21.1% of the total energy consumed.

*3. Densification Equipment*

***a. Process Considerations***

Among the many types of densifiers, the two most widely used in waste-to-energy facilities are the cubetters and pellet mills. An example of the pellet mill type is a unit manufactured by the California Pellet Mill Company (San Francisco, Calif.). The pellet mill consists of a rotating ring type die into which the die holes have been drilled radially. The die is contained in a housing and is driven by a motor. Also within the housing are two rolls installed within the inner portion of the die. The rolls perform the extrusion. They are not directly driven but are rotated by means of contact (friction) between them and the die. Material to be densified is introduced into the housing and is

**Table 4**
**FUEL CHARACTERISTICS OF TWO SAMPLES OF DENSIFIED RDF**

| Item | As received A % | As received B % | Moisture free A % | Moisture free B % |
|---|---|---|---|---|
| Moisture | 27.9 | 27.6 | — | — |
| Volatile matter | 54.9 | 51.3 | 76.0 | 70.6 |
| Ash | 10.2 | 11.9 | 14.2 | 16.6 |
| Hydrogen | 4.2 | 3.8 | 5.8 | 5.2 |
| Carbon | 29.0 | 34.4 | 40.2 | 41.8 |
| Nitrogen | 0.3 | 0.3 | 0.4 | 0.4 |
| Oxygen | 28.3 | 26.0 | 39.1 | 35.8 |
| Sulfur | 0.2 | 0.2 | 0.3 | 0.3 |
| Heating value (J/g)[b] | 11,255 | 11,879 | 15,657 | 16,378 |

Modified from Trezek, G. J., Diaz, L. F., Savage, G. M., and White, R., Prediction of the Impact of Screening on Refuse-Derived Fuel Quality, prepared for Electric Power Research Institute, Report No. EPRI FP-1249, November 1979.

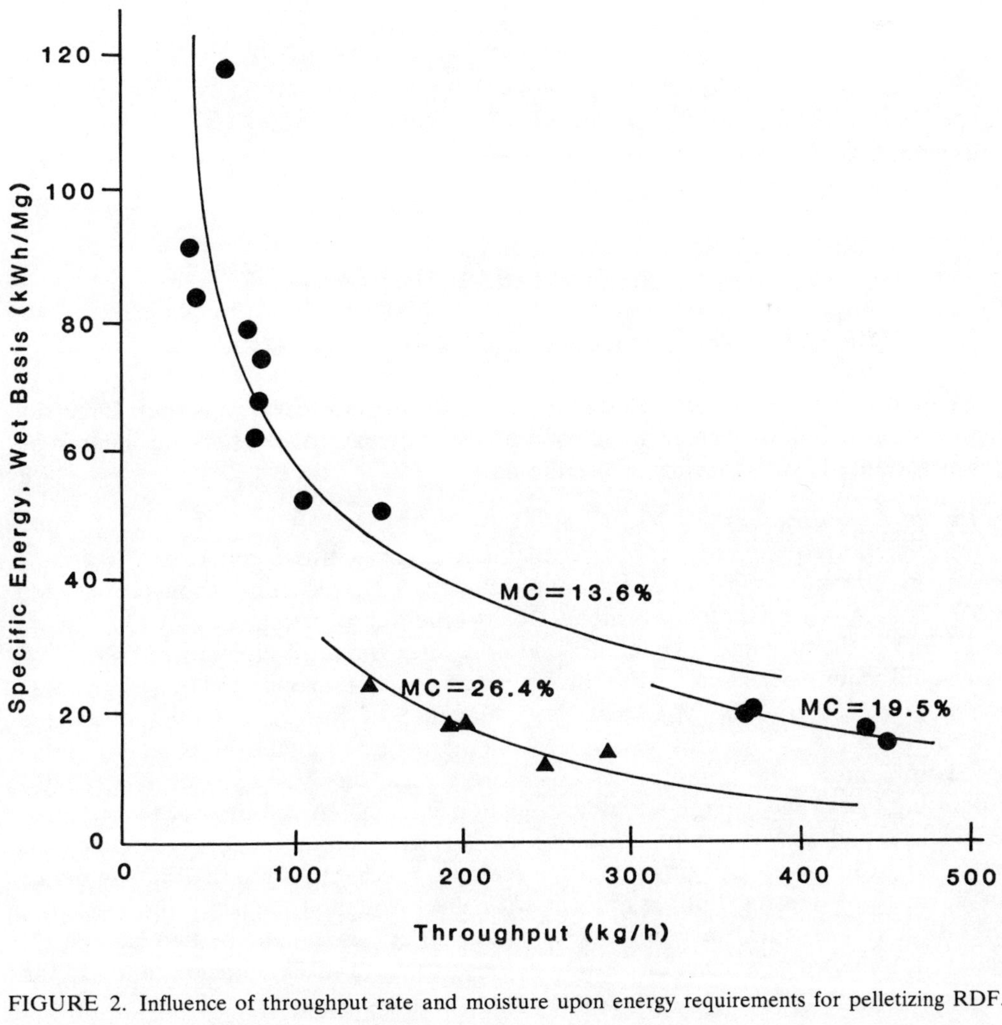


FIGURE 2. Influence of throughput rate and moisture upon energy requirements for pelletizing RDF.

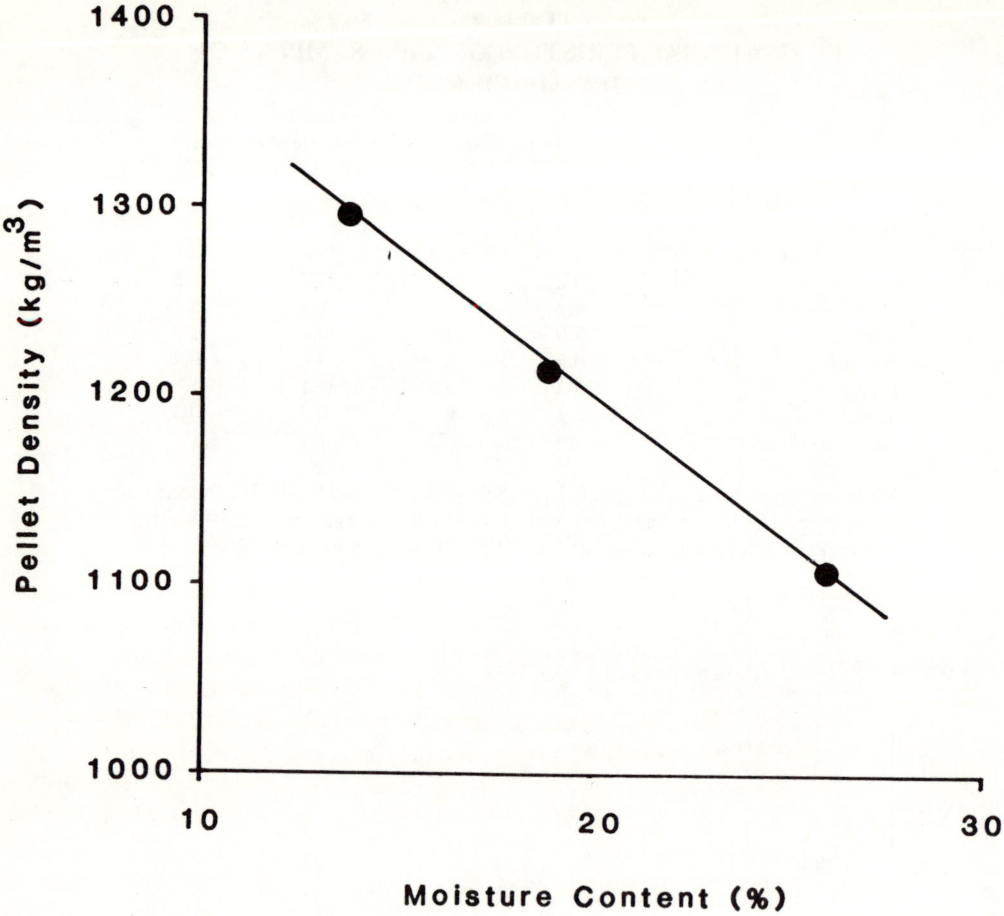


FIGURE 3. Influence of moisture content on pellet density.

distributed across the inside face of the die. The friction-driven rolls then force the contents through the openings in the rotating die. The extruded pellets are sheared by knives mounted on the inside of the die cover.

***b. Design Considerations***

Densification equipment presently available (e.g., briquetters, pelletizers, "cubers") apparently are impractical for all but small-scale resource recovery systems (< 150 Mg MSW/day). A principal reason is the relatively limited processing capacity (3 to 10 Mg dRDF/hr) of the machines in comparison to the large quantities (300 to 1500 Mg/day) that would require densification in a resource recovery system operated by a large urban community (600 to 2000 Mg MSW/day). Because of the limited capacity of the machines, it often becomes necessary to employ several densifiers and parallel processing lines. The complexities associated with the employment of multiple processing lines render large-scale densification systems, at the time of this writing, technically and economically infeasible as far as resource (energy) recovery is concerned. Moreover, bringing about the scale-up needed to more realistically match the capacity of densification equipment with the throughput capacity required of a large-scale resource recovery operation is made all but impossible by two factors: (1) the extrusion hardware must be extremely hard so as to resist abrasion; and (2) the geometry of the hardware precludes scale-up.

**Table 5**
**APPROXIMATE ENERGY REQUIREMENTS FOR THE PRODUCTION OF D-RDF**

| Unit process | Specific energy (k/Wh/Mg of MSW) | Percent of total |
|---|---|---|
| Size reduction | 18.8 | 56.6 |
| Air classification | 4.4 | 13.3 |
| Densification[a] | 7.0 | 21.1 |
| Miscellaneous equipment | 3.0 | 9.0 |
| Total | 33.2 | 100.0 |

[a] Assuming an air classifier recovery of 70% lights and 30% heavies.

Based upon Trezek, G. J., Diaz, L. F., Savage, G. M., and White, R., Prediction of the Impact of Screening on Refuse-Derived Fuel Quality, prepared for Electric Power Research Institute, Report No. EPRI FP-1249, November 1979.

While the problem of limited capacity is by far the greatest impediment to the application of densification in resource recovery from MSW, it is by no means the only one. Materials handling is a source of several difficulties which mainly take the form of material bridging or jamming in the equipment. The problems are especially encountered in the use of equipment designed for the agricultural industry without first modifying it to accommodate MSW. This latter problem is applicable to most of the commercially available densifiers. Agricultural densifiers are designed to process livestock feed and crop wastes. Needless to say, the handling characteristics of the two are a far cry from those of MSW.

A very severe drawback inherent in densification is the need to size reduce the input material to an exceedingly fine particle size. Pelletizers and "cubers" typically require a finely sized feed in order to produce an acceptable densified product. Moreover, they must do so within limits of energy consumption and die wear that are economically and practically feasible. The need for size reduction stems from the fact that densifiers are very inefficient size reduction devices, and they can successfully densify only that material which has a nominal particle size on the order of 1 to 4 cm. Typically, at least two stages of shredding are required. Therefore, the cost of shredding the feed waste to a very fine particle size must be added to that due to the intensification of the wear on the densifier brought about by the densification of the light fraction.

Although the state of the technology at the time of this writing may not have made densification advisable for large-scale resource recovery processing, the prospects for success with small-scale operations (150 Mg MSW/day or less) were and are very favorable. The small-scale systems can be designed to fit within the capacity of a bank of two or three of the largest commercially available densifiers operated in parallel.

## C. Utilization of RDF

Refuse-derived fuel has been used with success as a supplement to coal in a stoker-fired furnace and somewhat less successfully in suspension burning.[11-14] To accomplish the desired complete combustion in a suspension burner, it is necessary to reduce the nominal particle size of the RDF to 1.25 cm or less. This can be done through primary size reduction or through a secondary shredding. On the other hand, the refuse light fraction as produced, or as pelletized, can be satisfactorily burned in a stoker-fired furnace.

*1. Difficulties*

As of this writing, despite a history of successful utilization of RDF, albeit short, industry has been very reluctant to burn the fuel in conventional steam generation units. Among the reasons for the reluctance, five stand out: (1) the varied particle size and composition of most refuse-derived fuels; (2) the presence of materials (i.e., "contaminants") that are characterized by unsatisfactory combustion properties; (3) high moisture content; (4) relatively high ash content; and (5) a mediocre heating value as compared to conventional fuels.

*2. Variable Composition*

Even though the light fraction is a much more homogeneous material than is raw refuse, nevertheless it cannot be considered a pure material from the standpoint of combustion. Because the light fraction is a mixture of materials, it is difficult to maintain an efficient combustion process when it serves as the fuel. Material and moisture content variations in the light fraction may result in daily fluctuations in energy content by as much as ±15%.

*3. Presence of Contaminants*

Although the light fraction of MSW consists of an abundance of paper (heat value, 17,400 J/g) and plastic (heat value, 37,100 J/g), it also includes materials, the burning of which results in the production of large amounts of ash and substances potentially damaging to burner and boiler structures, and detrimental to the quality of the exhaust gases. For example, during combustion, the chlorine in certain plastics (e.g., PVC) and in sodium chloride reacts with water vapor to form hydrogen chloride (HCl). In the presence of water vapor, the HCl forms the highly reactive hydrochloric acid, which in turn actively attacks the metallic tubes and other heat transfer surfaces in the burners and boilers.[15]

*4. High Moisture Content*

The high moisture content of refuse-derived fuel detracts from its quality as a fuel in three ways:

1. Since water has no inherent heating value, and therefore acts essentially as an inert compound in terms of fuel value, the presence of the moisture serves to reduce the overall heating value of the RDF.
2. Upon evaporation, moisture acts as an energy absorber during the combustion process. In effect, the moisture "steals" useable heat from any energy recovery process that may be downstream.
3. Beyond the zone of active combustion, the moisture may recondense on cool surfaces of the burner or heat exchange equipment. The condensed water can then react with gases in the emissions from the combustion process and form corrosive compounds.

A more detailed explanation of the mechanisms and principles involved is given in the paragraphs that follow.

If RDF were burned in a typical solid fuel-fired steam generation unit, the contained moisture would be heated, evaporated, superheated, and subsequently exhausted as water vapor along with the combustion gases. Energy thus absorbed by the water normally cannot be recovered because the temperature of the exhaust gas usually is maintained above dew point so as to prevent condensation from taking place in the boiler section. The enthalpy function can be used as a convenient quantity for determining the energy absorbed by water or water vapor due to heating. If heat is added to a substance at constant pressure, as is the case in a furnace, no external work is done by the system, and all heat is used to increase the enthalpy of the substance. Consequently, the change in enthalpy represents the heat absorbed at constant pressure. This approach is illustrated in the description presented in the two following paragraphs. In the description, the only concern is with the energy loss resulting from the moisture content in the fuel. Hence, the additional water formed in the combustion reaction ($CH_2O + O_2 \rightarrow CO_2 + H_2O$) is not taken into consideration. Furthermore, the heat transfer efficiency of the boiler is assumed to be constant (80%) with each type of fuel, and accordingly, it is neglected.

For the purpose of illustration, it is assumed that the temperature of the screened and unscreened light fractions is 21°C (70°F), and that the furnace is operating at atmospheric pressure ($1 \times 10^5$ N/m$^2$). Under those circumstances, the moisture in the fuel absorbs energy as it is heated from 21 to 100°C (212°F). Upon reaching 100°C, the moisture evaporates and the water vapor is subsequently superheated to the equilibrium temperature of the combustion gases. At this point another assumption is made, namely, that the gases (including the water vapor) are cooled to 121°C (250°F) after passing through the boiler. In the boiler, heat from the combustion gases is transferred to the water flowing through the boiler tubes.

Heat lost in the exhausted water vapor (i.e., not recovered in the boiler) can be expressed on the basis of energy per unit weight of water. The unrecovered heat is in three forms: (1) the energy required to heat the water from 21 to 100°C at $1 \times 10^5$ N/m$^2$ (330 J/g $H_2O$ or 142 Btu/lb); (2) the heat of vaporization at 100°C and 1.0 atm (2251 J/g or 970 Btu/lb); and (3) the energy required to superheat the water vapor from 100 to 121°C (assuming a constant pressure of $1 \times 10^5$ N/m$^2$ (1.0 atm)), which amounts to 44 J/g (19 Btu/lb). The total heat loss is the sum of the three forms, namely, 2625 J/g (1,131 Btu/lb) of water.

*5. High Ash Content*

Inasmuch as RDF is a solid fuel, it leaves an ash residue after combustion. Therefore, if the RDF is to be used in a furnace designed to burn gas or oil, provision would have to be made for removing the ash. On the other hand, since coal burning furnaces are equipped to handle ash, RDF can be used as a fuel in them without incurring an ash removal problem. Inasmuch as coal and RDF are solid fuels that can be burned in similar equipment, in the remaining discussion on RDF its fuel value is considered in terms of its similarity to coal as a fuel.

**D. Mediocre Heating Value**

The mediocrity of RDF with respect to coal in terms of comparative heating values and of moisture content and ash is well illustrated by the data in Table 6. Although the data pertain to RDF and coal used in the St. Louis project, they are representative of the two fuels in general. Since the data are representative, it can be assumed that the heating value of coal is more than twice that of the RDF. On the other hand, data in the table show that the moisture content of RDF can be expected to be twice that of coal, and the ash content almost three times as much. Clearly then, from the standpoint of combustion, the combustion properties of RDF leave something to be desired. Because of the great disparity between the heating value of RDF and that of coal, two times as much RDF as coal must be burned to yield a given amount of energy. For example, on an as-received basis, 84 g of RDF must be burned to yield $1 \times 10^6$ J of energy; whereas the same yield could be obtained by burning only 37 g Orient coal. On the other hand, the amount of ash produced from the combustion of the RDF would be about 19 g,

**Table 6**
**FUEL CHARACTERISTICS OF REFUSE-DERIVED FUELS AND ORIENT COAL**

| | Heating value | | Moisture | Ash | | |
|---|---|---|---|---|---|---|
| | | | | | Production | |
| Fuel | J/g | Btu/lb | (%) | Content (%) | g/MJ | lb/MBtu |
| RDF | | | | | | |
| Unscreened | 12,111 | 5,218 | 23.1 | 20.9 | 17.3 | 40.0 |
| Screened | 15,541 | 6,696 | 16.3 | 12.1 | 7.8 | 18.1 |
| Coal ("Orient") | 26,875 | 11,579 | 12.5 | 7.6 | 2.8 | 6.6 |

Modified from Savage, G. M., Diaz, L. F., and Trezek, G. J., *Waste Age,* April 1978.

while that from the coal would be slightly less than 5 g. A practical consequence of the two differences is a magnification of existing storage and materials handling problems that would arise from a switch from firing coal to firing RDF in a furnace designed solely to burn coal. If the RDF were to replace gas or fuel oil, the resulting problem would be even greater, inasmuch as the heating values of the latter two are on the order of 44,100 J/g and practically no ash is formed in their combustion.

## III. IMPROVING THE FUEL QUALITY OF RDF

### A. Changes Required

Several characteristics of the light fraction of shredded refuse must be altered in order to bring about a satisfactory improvement in its fuel quality with respect to combustion, efficiency, and material handling. Although the characteristics in question were discussed in the preceding pages, they are named again at this point for purposes of emphasis. They are (1) low heating value as compared to coal, natural gas, and fuel oil; (2) a relatively high moisture content; and (3) a sizable ash content.

Research[4,16] has shown that the three more seriously objectionable characteristics of the light fraction, and hence of RDF, can be ameliorated to a significant degree through the addition of a relatively simple step in the overall RDF production process, namely, passage through a rotating cylindrical screen (trommel). Indeed, the improvements discussed in detail in the two sections that follow are almost entirely brought about through trommel screening.

### B. Removal of Fines

A major improvement in the fuel quality of refuse lights as brought about by trommel screening after air classification is the removal of inert contaminants present in the form of fines. According to results reported by Savage et al.,[5] more than 90% of the materials of low heating value present in the light fraction is in the form of inert fines, and are removed through trommeling. Moreover, the paper content of the light fraction after trommeling is 85% as compared to its original 60%. In general, the oversize material retained on the screen, i.e., screened lights or RDF, represents from 40 to 60% of the raw refuse as discharged from the collection vehicle.

The improvement in heating value is impressive. For example, the heating value of trommel-screened lights averages about 18,570 J/g (8000 Btu/lb) of dry solids. On the other hand, the heating value of unscreened RDF produced in the St. Louis operation averaged 15,630 J/g (6730 Btu/lb). On the basis of these two sets of numbers, screening can be assumed to increase the heating value of the lights by about 20%.

The increase in heating value is not entirely explained by the removal of the inert fines. Other material, also removed in the screening process, may not necessarily be inert, nor in a form as minute as fines. The presence of this second group of materials was demonstrated in the study reported by Savage et al.[5] In the study, it was found that the air classified light fraction of refuse can be divided into two distinct parts on the basis of heating value, namely, a part characterized by a high-heating value, and a part characterized by a low heating value. The part characterized by high heating value consists mainly of paper and plastic. Its dry weight heating value is on the order of 18,600 J/g (8000 Btu/lb). The fraction having the low heating value is composed mostly of fines and organic (putrescible) materials. Its heating value averages about 10,800 J/g (4700 Btu/lb). Obviously, it is the inclusion of the latter material in the unscreened light fraction that reduces the overall heating value of the lights.

### C. Lowering of Moisture Content

The fuel value of the light fraction is further enhanced in screening through the resulting lowering of the moisture content. Typically, unscreened lights have a moisture content of 20 to 25%, whereas the screened product has a moisture content of only 15 or 16%. While the actual values of the two materials may differ, the relationship remains the same. The difference between the moisture contents is due to the fact that the lights that have the higher moisture content are separated from those that have the lower moisture content. The moisture content of the material that passes through the screen ("undersize" particles) is greater than that of the material retained on the screen ("oversize" particles). The "oversize" particles constitute the screened lights or "screened RDF". The effect of lowering the moisture content through trommeling is illustrated by the data in Tables 7 and 8.

By expressing the heat loss in terms of unrecovered energy per unit weight of dry solids, it is possible to make certain comparisons, one of which is as follows: if unscreened RDF has a moisture content of 22.5%, its water-to-solids ratio is 0.290. At such a ratio, the heat lost in the exhaust due to moisture in the fuel is 2625 J/g water × 0.29, which amounts to 761 J/g solids (328 Btu/lb). On the other hand, if the screened RDF (light fraction) has its usual moisture content (viz. 16.3%), the water-to-solids ratio is 0.195, and the energy lost because of moisture in the fuel is 513 J/g solids (221 Btu/lb). Another way of making the comparison is by expressing the unrecovered energy in terms of percent of the energy content of screened and unscreened RDF. With the energy content of the unscreened fraction at 12,111 J/g (5218 Btu/lb), the energy not recovered as heat in combustion is 4.9%. At a heating value of 15,541 J/g (6696 Btu/lb), the percent energy not recovered from screened RDF is 2.8

## IV. PROBLEMS AWAITING SOLUTION

### A. Firing in Utility Boilers

Despite the progress made in improving the fuel characteristics of RDF through trommel screening, difficulties remain to be resolved. It can be expected that the use of RDF in a utility boiler will have a definite impact on the performance of the boiler, of the electrostatic precipitator, of the ash removal system, and on the operation and maintenance costs of the plant.

A major problem, or at least a cause for concern, is the matter of air pollution resulting from the combustion of RDF. It is true that because of the low sulfur content (<0.15%) of RDF, the emission of oxides of sulfur should not constitute a problem. On the other hand, there may be cause for concern with respect to particulates, carbon monoxide and hydrogen chloride discharges, regardless of whether or not the burned RDF has been screened.

**Table 7**
**EFFECT OF EXTENT OF MOISTURE CONTENT ON HEATING VALUE**

| Item | Light fraction | | | |
|---|---|---|---|---|
| | Not screened | | Screened | |
| Moisture content | 22.5 | | 16.3 | |
| Heating value | | | | |
| Dry basis, J/g (Btu/lb) | 15,627 | (6733) | 18,568 | (8000) |
| As received, J/g (Btu/lb) | 12,111 | (5218) | 15,541 | (6696) |
| Decrease in heating value | 22.5 | | 16.3 | |

Modified from Savage, G. M., Diaz, L. F., and Trezek, G. J., *Waste Age,* April 1978.

**Table 8**
**ENERGY LOSS BECAUSE OF UNRECOVERABLE HEAT ASSOCIATED WITH MOISTURE CONTENT VAPORIZATION**

| Item | RDF | |
|---|---|---|
| | Unscreened | Screened |
| Heating value (J/g)[a] | | |
| Dry basis | 15,627 | 18,568 |
| As is (wet) | 12,111 | 15,541 |
| Energy lost in exhausted $H_2O$ vapor (J/g)[a,b] | | |
| As-received basis | 590 | 427 |
| Dry basis | 761 | 513 |
| Net recoverable energy | | |
| As received (J/g)[a] | 11,521 | 15,114 |
| Moisture content (%) | 22.5 | 16.3 |

[a] To convert to Btu/lb, multiply by 0.4296.
[b] Energy loss per gram $H_2O$ vaporized, 2625 J/g.

Based upon Savage, G. M., Diaz, L. F., and Trezek, G. J., *Waste Age,* April 1978.

As stated earlier, the relatively high moisture and ash contents of RDF result in a reduction in power output (load), in that the moisture leads to an increase in gas volume and the ash brings about slagging. The reduction amounts to 1 MW/Mg/hr of RDF. To control the slag formation it is necessary to maintain higher-than-normal concentrations of $O_2$ in the furnace. Another objectionable consequence of the slag is a lessening of the amount of heat transfer accomplished in the fire box. As a result, the temperature of the gas discharge is raised.

Tests conducted in facilities in which RDF is used have shown that the amount of particulates increases with increase in the percentage of RDF fired.[17] Moreover, the efficiency of the electrostatic precipitator decreases in proportion to the increase in the contribution to the heat input by the RDF.

The consensus is that the burning of RDF adversely affects an ash handling system to a serious extent, if for no other reason than that the amount of ash to be disposed is increased. The problem is compounded by the fact that much of the RDF falls through the fire ball (zone) without being burned. The unburned material consists of large pieces of wood and hard plastic, textiles, etc. Since some of the unburned items are buoyant, they also adversely affect the wastewater treatment system. The problem of unburned particles can be substantially reduced through careful control of the particle size of the RDF.

*1. Ash*

Because of the serious impact had by ash on the performance of a utility boiler, the subject of the increased ash production associated with RDF combustion is further explored at this point.

Refuse-derived fuel and coal contain incombustible (inorganic) matter in amounts conducive to the development of numerous problems in a large utility boiler. Obviously, being incombustible, the material is found in the ash. If the boiler is to be operated continuously, the ash must be removed. Ash is removed from stoker-fired furnaces by periodically shaking or dumping sections of grates, or by continuously moving the ash to a discharge point. In suspension firing, ash settles on various surfaces of the boiler. The ash can be removed by cleaning the boiler periodically. With both types of firing, i.e., stoker and suspension, the ash becomes molten if the furnace temperature is sufficiently high, and in that state can be removed continuously. However, given the proper conditions, the ash forms slag deposits on the furnace walls (slagging). Ash particles small enough to be carried by the gaseous stream may form deposits on the tubes (fouling).

Slags can cause severe corrosion problems in a furnace. Furthermore, because of their low thermal conductivity, an accumulation of the material on walls and tubes seriously interferes with heat transfer.

Viscosity and temperature play a major role in the accumulation of slag. The critical temperature level is that at which the flow of the molten ash changes from Newtonian to plastic, i.e., the Temperature of Critical Viscosity (TCV). Immediately upon the temperature falling below the TCV, the slag ceases to flow and begins to solidify. This latter results in an accumulation of slag.

The ash burden, (kg ash per MJ or lb ash per MBtu) affects slagging and fouling. The effect is due primarily to the fact that firing a fuel with a high ash content is attended by a corresponding increase in concentration of fly ash. As the amount of fly ash increases, the probability for collisions between ash particles and for subsequent formation of molten slag also increases. To minimize the problem, numerous attempts are being made to reduce coal and RDF ash by physical or chemical means, or by a combination of the two.[4, 16, 18-20]

**B. Potential Solutions**

Experience in waste processing and RDF firing shows that a number of the problems discussed in the preceding section can be solved through the attainment of:

- A better understanding of the characteristics of MSW and RDF
- A more complete knowledge of operating parameters of unit processes
- A better control of particle size
- An improvement in the quality of RDF
- An optimum sequence of unit processes to meet boiler requirements
- An understanding of principles of RDF firing
- The design of new units intended specifically to burn RDF or mixtures of RDF and fossil fuel (see Figure 4)

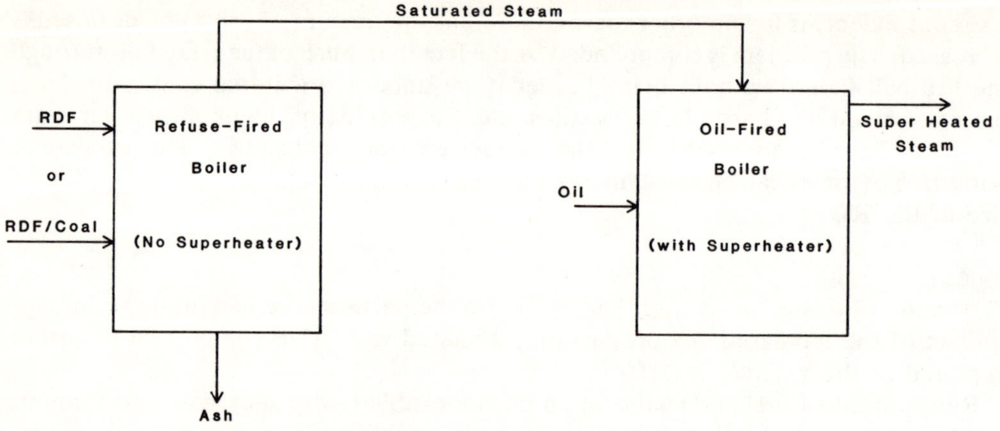


FIGURE 4. Conceptual design of dedicated refuse/coal/oil firing system.

## V. RDF FACILITIES

### A. St. Louis Demonstration Plant

Even though the St. Louis facility is no longer in operation, a description of the plant is given herein because the operation exemplified the evolution of RDF production into a modern practice. Incidentally, because the plant is not an ongoing one, verbs used in describing are either in the past or in the perfect tense.

#### *1. Waste Processing*

Municipal solid waste was delivered to the plant in conventional collection vehicles. The sequence of the steps in the operation of the plant is self-evident from the flow diagram shown in Figure 5. The shredder was a horizontal shaft hammermill powered by a direct drive, 932 kW (1250 hp), 900 rpm motor. Following air classification, the light fraction of the shredded refuse was transported to a storage bin. The heavy fraction was passed by a magnetic separator to recover the ferrous scrap. The ferrous scrap was processed through a 75-kW (100 hp) "nuggetizer", and the nonferrous heavies were transported to a land fill.

After having been removed from the storage bin, the lights were compacted into a 57-$m^3$ (75 cubic yard) self-unloading transfer trailer and were transported approximately 32 km (20 miles) to the Union Electric Plant at Meramec, Missouri.

#### *2. Receiving and Firing Facilities*

The recovery facilities at Meramec are diagrammed in Figure 6. The receiving building, although located at the power plant, was built and operated by the city of St. Louis as part of the processing system. The RDF was dumped from the transport truck into a receiving pit (capacity 18 to 23 Mg) from which it was pneumatically transferred to an Atlas® bin. This transfer was accomplished in part by means of a double-auger screw conveyor which slowly traversed the bottom of the pit to pick up and feed the RDF onto a belt conveyor. The RDF was discharged from the belt conveyor through an airlock and into a 30-cm pneumatic conveying line through which the material was conveyed to the Atlas® bin (capacity 54 Mg). The time needed to transfer a truckload of RDF from the receiving pit to the Atlas® bin was about 1 hr. A pneumatic feeder system conveyed the RDF from the Altas® bin through four separate pipelines directly into firing ports in each corner of the furnace.

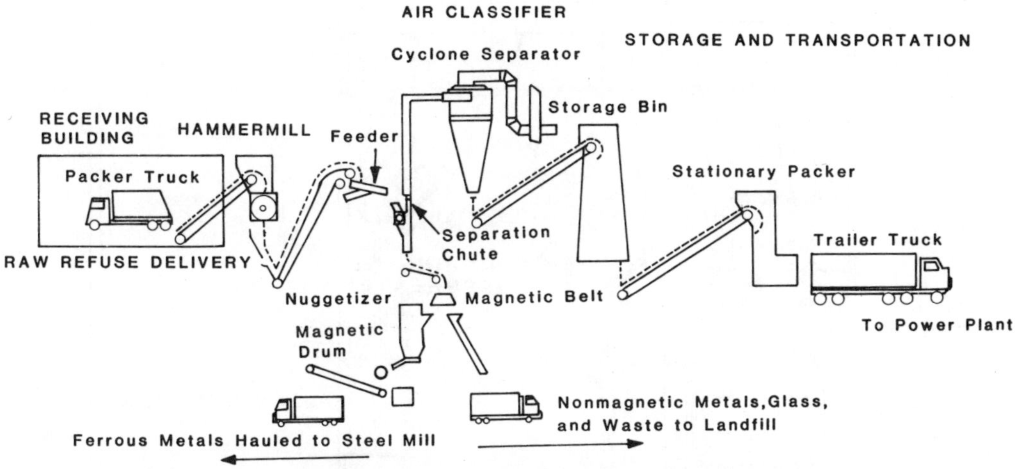


FIGURE 5. Flow diagram of refuse processing facility at the St. Louis plant.

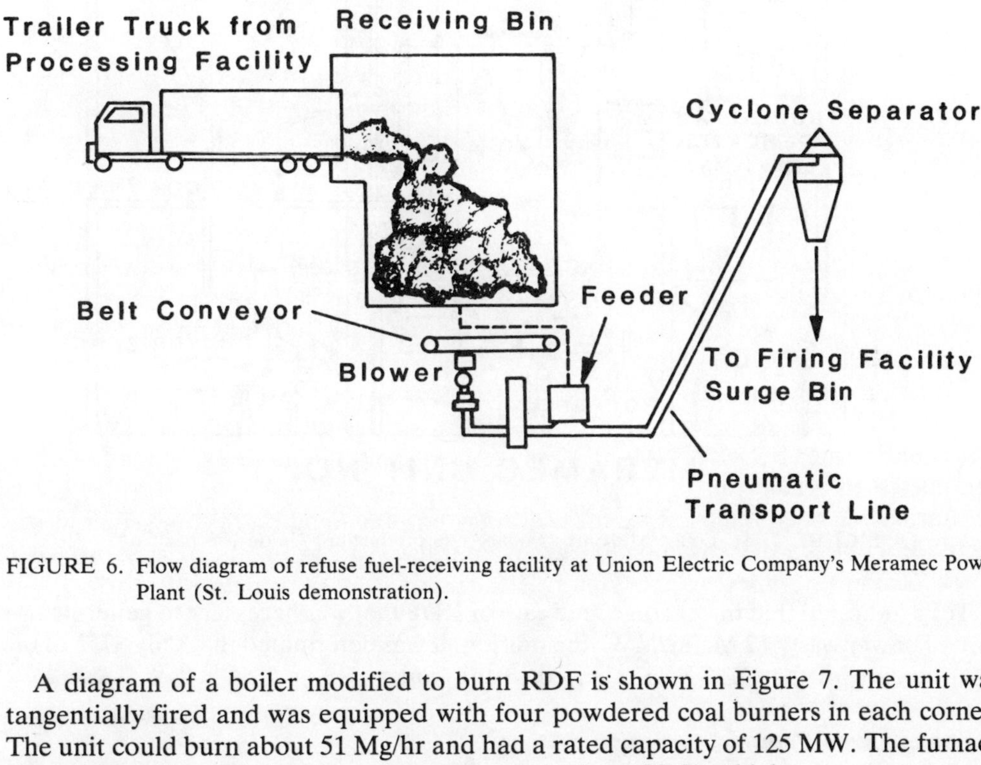


FIGURE 6. Flow diagram of refuse fuel-receiving facility at Union Electric Company's Meramec Power Plant (St. Louis demonstration).

A diagram of a boiler modified to burn RDF is shown in Figure 7. The unit was tangentially fired and was equipped with four powdered coal burners in each corner. The unit could burn about 51 Mg/hr and had a rated capacity of 125 MW. The furnace was about 8.5 × 11.5 m in cross-section and about 30.5 m high.

The RDF nozzles were located between the two upper and two lower coal nozzles. Sufficient velocity was imparted to the particles to carry them into the furnace high-temperature zones where the particles were ignited and burned. Light particles were carried out with the flue gas, and heavy unburned RDF particles fell into the boiler ash pit.

The RDF firing system was completely independent of the main combustion control system. The boiler operator could only initiate or stop RDF firing. The firing rate could be adjusted manually only at the control center of the RDF bin motor.

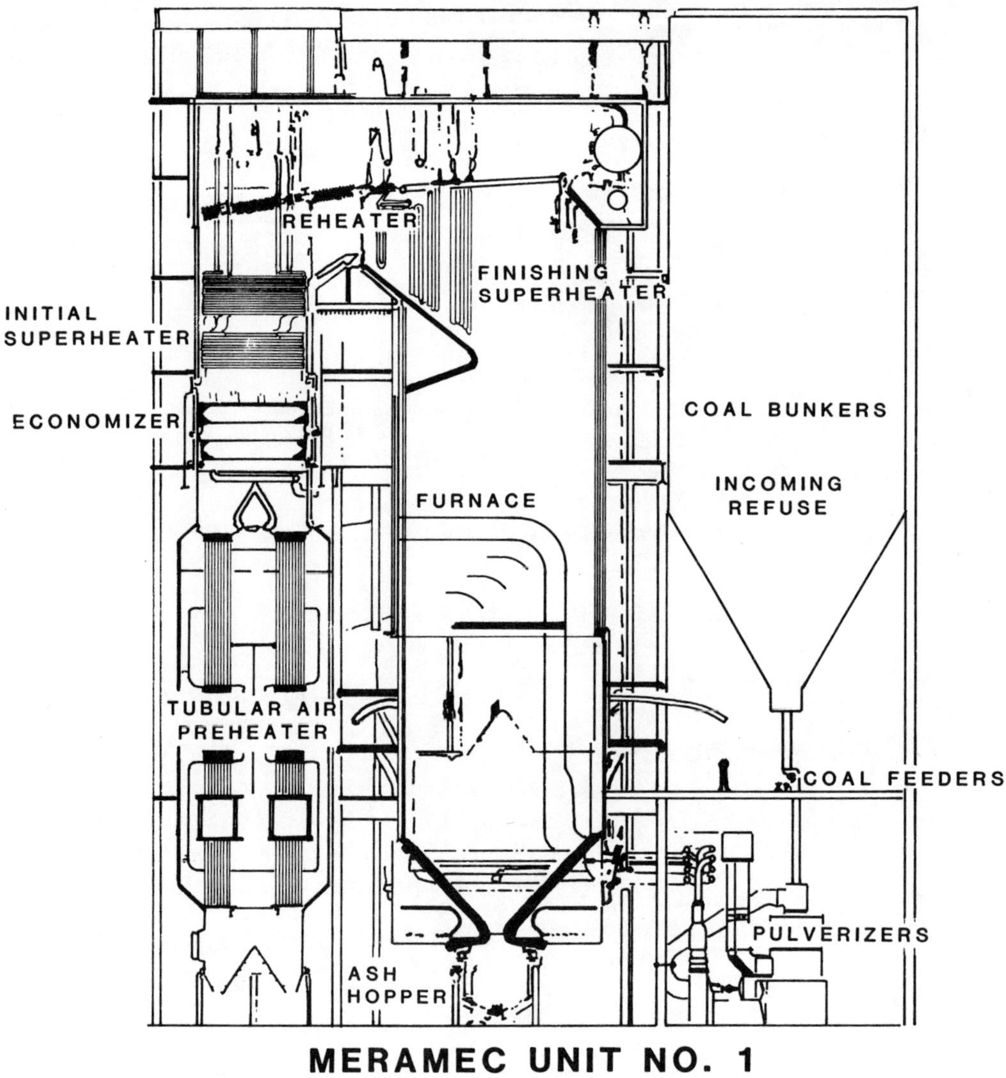


FIGURE 7. St. Louis, Missouri auxiliary fuel production, Union Electric Co.

Tests indicated that the average feed rate of RDF that was necessary to generate one unit of power was 1.12 Mg/hr/MW. In addition, it was determined that only 87% of the heat content of the RDF was released in the boiler.

## B. Ames, Iowa Plant

### *1. Refuse Processing*

Operations were begun in the Ames, Iowa recovery plant in 1975. The waste processing facility of the plant is designed to recover recyclable materials and energy from MSW at a nominal capacity of 136 Mg/day.

In the operation, the wastes are weighed and then transferred to a 746-kW (1000 hp) shredder to be size reduced. The shredded material is exposed to a magnet to remove ferrous wastes. Thereafter, it passed through a second shredder (746 kW.) The shredded refuse is conveyed to a surge bin, and eventually, from the bin into an air classifier. The light fraction (lights) is transported pneumatically into an Atlas® bin, and the heavy fraction is subjected to magnetic separation. Although the plant was designed to provide aluminum and glass recovery, thus far, its performance in this respect has left much to be desired.

In 1978, the plant was subjected to major modifications, one of which was the installation of the grit removal system diagrammed in Figure 8. The key items in the modified system are the two disc screens.[18]

*2. Firing Facilities*

The RDF produced in the refuse processing plant can be burned in three of the units at the Ames Municipal Power Plant. Two of the three units are stoker fired, and the third is suspension fired. Characteristics of the three units are listed in Table 9.

The initial design of the system called for burning RDF as a supplementary fuel with coal in the suspension-fired unit, and for RDF to be burned in the stoker-fired units only when suspension-fired unit was shut down. However, that plan could not be followed, because in the suspension-fired unit a large amount of combustible matter dropped through the fire zone unburned and accumulated at the bottom of the furnace. Because of this malfunction, continuous firing became an impossibility. The problem was rectified by the installation of grates in the unit.

**C. Wisconsin Electric Power Company (Milwaukee) Plant**

*1. Refuse Processing Plant*

All municipal solid waste collected in the city of Milwaukee is processed at the local Americology resource recovery plant. Inflow ranges from approximately 720 Mg/day in winter to 1600 Mg/day in summer.

A flow diagram of one of the two lines in the refuse processing plant is shown in Figure 9. Hand sorting is done to remove explosive substances from the throughput stream. The shredder is a 746-kW (1000 hp) Williams Patent Crusher® designed to reduce the particle size of the refuse to less than 13 cm. The zig-zag classifier indicated in the diagram was designed by Americology engineers. Ferrous metals are removed from the heavy fraction by Dings® magnetic separators, and the remaining material is land filled. Equipment installed to recover glass (trommel screen) and aluminum (eddy current separator) presently is not in operation.

The light fraction is collected in a cyclone collector and fed by a rotating airlock into a Heil® secondary shredder, in which approximately 90% of the fraction is reduced to less than 2 cm in size. Beyond this point, the processing involves passage of the lights through a cyclone, under a magnetic separator, and over Rader® disc screens to remove noncombustibles. The "lights" remaining after the noncombustibles are removed are compacted into 15-Mg transfer trailers and trucked approximately 32 km to the Wisconsin Electric Power Company (WEPCO) plant in Oak Creek, Wisconsin to be burned as fuel.

*2. Receiving and Firing Facilities*

The RDF receiving and storage facilities at WEPCO's Oak Creek plant are schematically diagrammed in Figure 10. The receiving area can accommodate three trucks simultaneously. The trucks discharge directly onto a 3.65-m wide conveyor which is sunken about 3.0 m below the floor surface. The discharged material is transported to a blower where it first falls upon a fluffing drum and then is passed through an airlock to be pneumatically conveyed to an Atlas® storage silo via a 41-cm-diameter pipeline.

Although the silo has a 812 Mg capacity, it is filled only to 450 Mg when RDF burning is taking place, and to 317 Mg when burning is not taking place. A rotating distributor at the top of the silo distributes the material evenly into a cone shaped pile. Four sweep

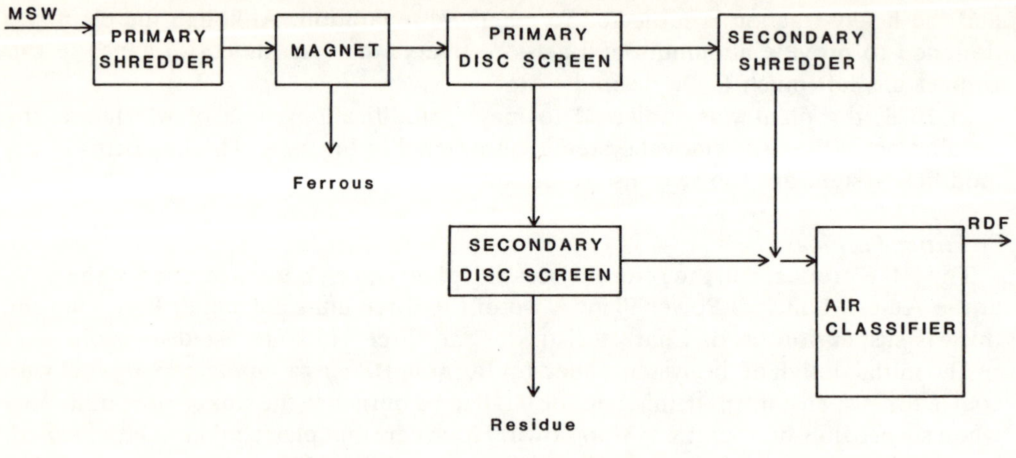


FIGURE 8. Flow diagram of the grit removal system at the Ames facility.

**Table 9**
**CHARACTERISTICS OF AMES MUNICIPAL POWER PLANT STEAM GENERATORS**

| Manufacturer | Riley (stoker fired) | Union iron works (stoker fired) | Combustion engineering (suspension fired) |
|---|---|---|---|
| Electrical output—MW | 7.5 | 12 | 35 |
| Installation date | 1951 | 1958 | |
| Pressure/temp | | | |
| pKa/°C | 4895/441 | 4999/441 | 5860/481 |
| (psi/°F) | (710/825) | (725/825) | (850/900) |
| Nominal steam output capacity | | | |
| kg/hr | 43,100 | 56,700 | 163,000 |
| (lb/hr) | (95,000) | (125,000) | (360,000) |
| Coal firing equipment | Spreader, stoker, traveling, grate | Spreader, stoker, traveling, grate | |
| Furnace pressure | Balanced, draft | Balanced, draft | |
| Dust collection equipment | Western, Multiple, Cyclone | American Blower, Multiple, Cyclone | American Standard electrostatic precipitator |
| Stack height | | | |
| Meters | 61 | 61 | 61 |
| (feet) | (200) | (200) | (200) |
| Heat input at nominal capacity | | | |
| MJ/hr | 154 | 202 | 469 |
| (Btu × $10^6$/hr) | (146) | (191) | 445 |

bucket strings at the base keep the material constantly moving, as well as feed the eight discharge conveyors located just below the floor level. Two discharge conveyors feed each of four blowers, dropping the material onto another fluffing drum and then on through an airlock to the blower. The blowers pneumatically convey the RDF to the boilers. A baghouse filters dust from the silo as well as the truck discharge area.

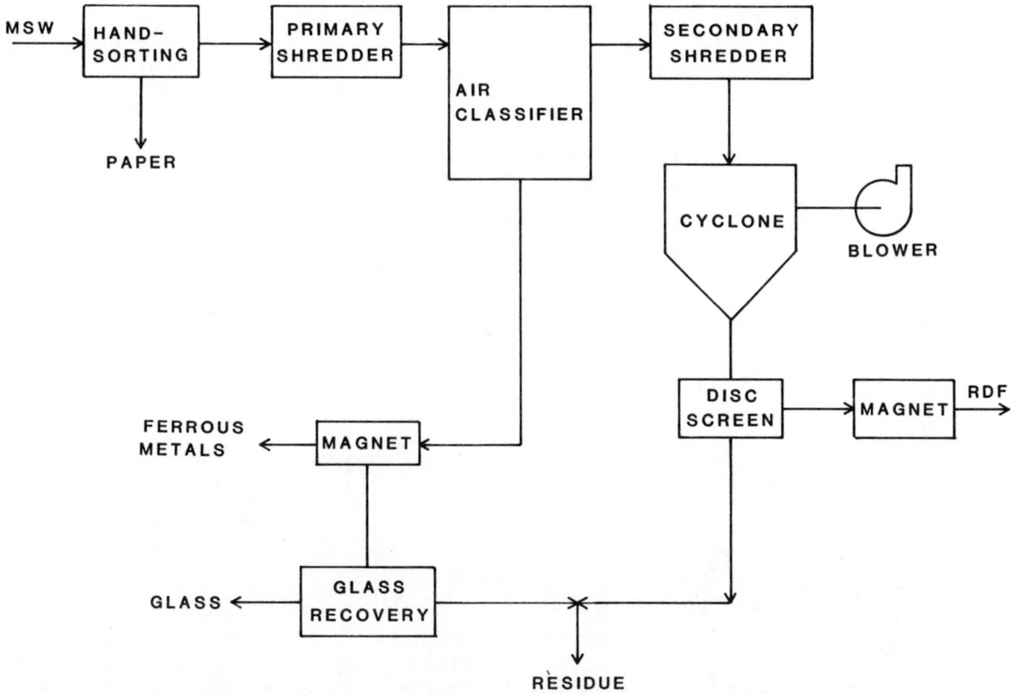


FIGURE 9. Flow diagram of the Milwaukee refuse processing site.

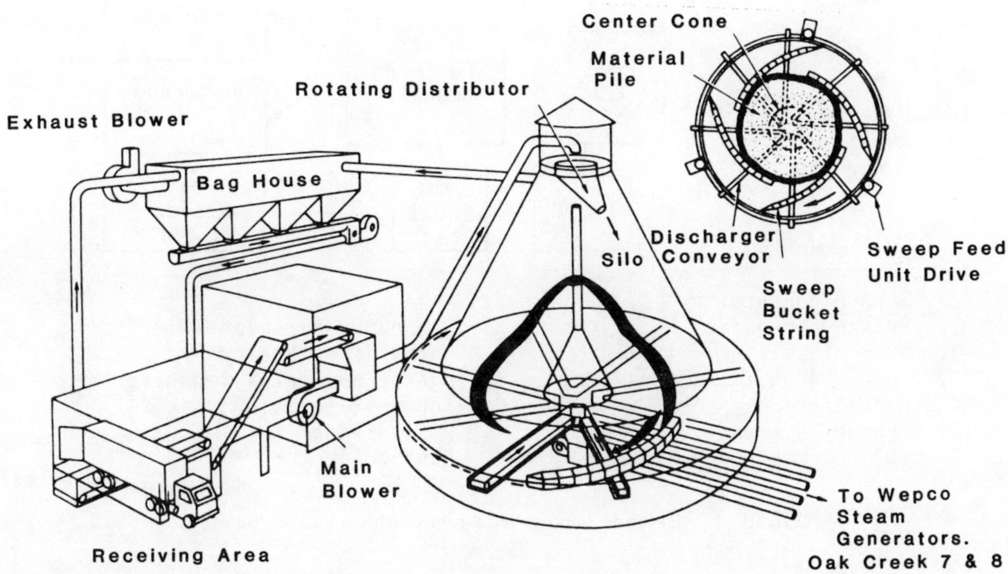


FIGURE 10. Schematic diagram of the receiving and storage facility of WEPCO, Oak Creek, Wisconsin.

Two modified pulverized coal boilers are being used to cofire the RDF with coal. The boilers are Combustion Engineering balanced-draft tangentially-fired units designed to operate at 16,400 kPa (2400 psig) and 571/535°C (1050/1000°F).

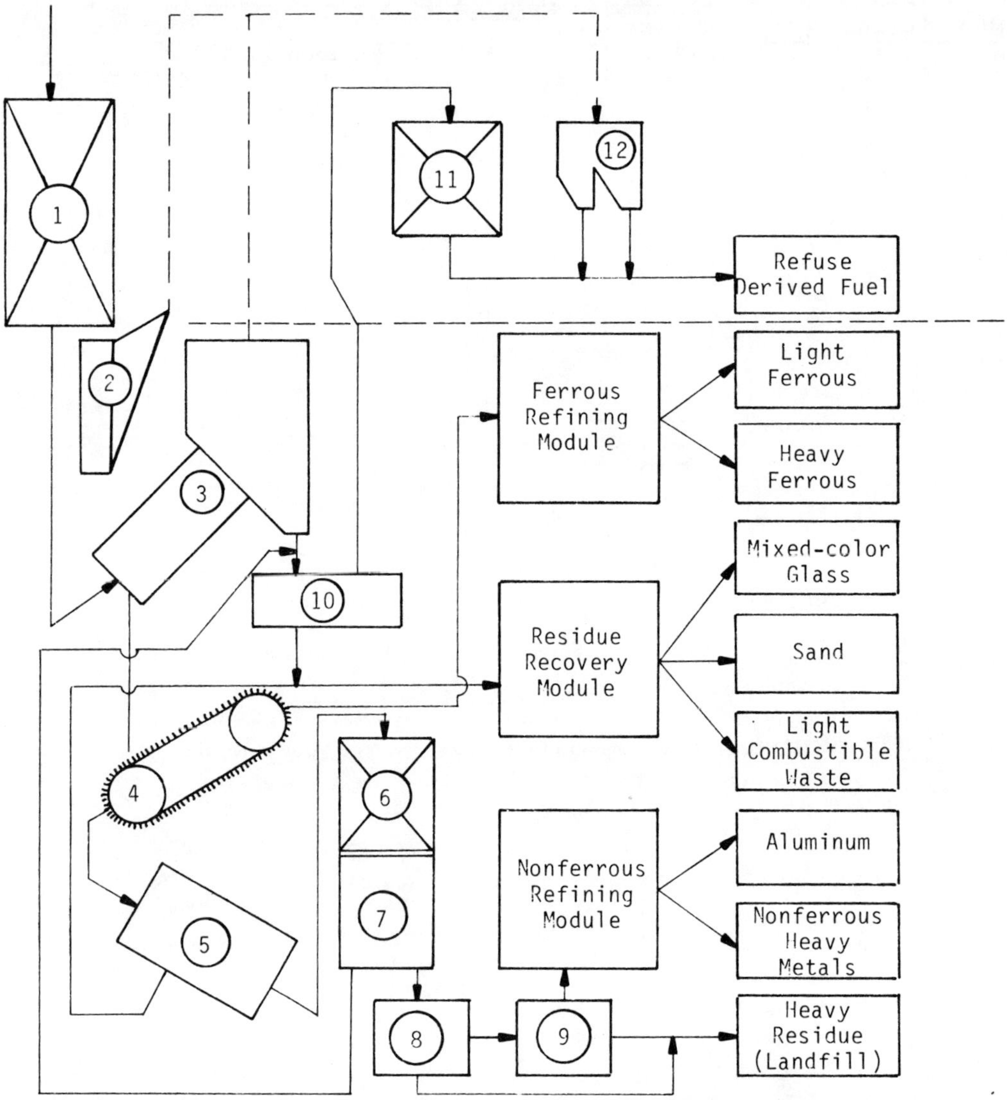


Key for Components

1. Primary shredder
2. Skim classifier
3. Rotary drum air classifier
4. Magnetic separator
5. Trommel screen, 19 mm holes
6. Secondary shredder
7. Secondary air classifier
8. Magnetic separator
9. Nonferrous metal separator
10. Screen (vibrating)
11. RDF shredder
12. Dust collector

FIGURE 11. Baseline process with relationship to other modules.

## D. Rochester Gas and Electric Company (New York)

### *1. Refuse Processing*

The recently completed Monroe County Resource Recovery Facility in Rochester, New York is designed to process 1800 Mg of MSW per day, and thereby to produce 1200 Mg of RDF per day. A flow diagram of the processing plant is shown in Figure 11. The

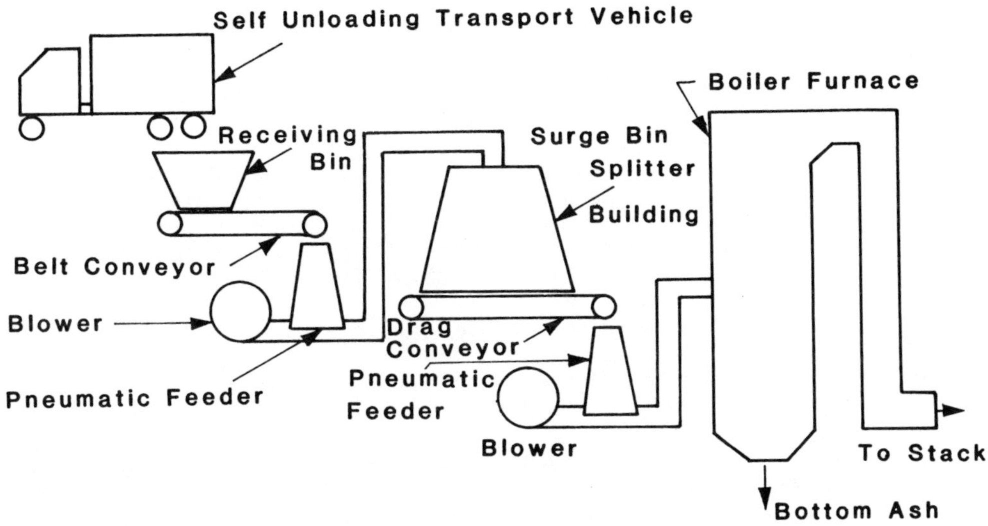


FIGURE 12. RDF receiving and handling facility at power plant.

primary shredders indicated in the figure are each driven by 746 kW (1000 hp) reversible motors, and are each rated at 63 Mg/hr. The particle size of the shredded product is less than 30 cm. Separation into "lights" and "heavies" is accomplished by means of four 31.5—Mg/hr rotary drum air classifiers—two for each shredder. A feature of the system is the return of the dust from the air cleaning system to the "lights", thereby making the dust a constituent of the RDF product.

The heavy fraction from the air classifiers is subjected to the series of processes indicated in the diagram in Figure 11. As for the ferrous material removed from the heavy fraction, it is further separated into a heavy portion suited for sale in the scrap iron market, and a light portion fit for sale in the higher priced, detinning market. The nonferrous heavies are passed through a trommel. The "undersize" from the trommel is subjected to the various steps indicated in Figure 11. The result is a separation of the "undersize" into glass cullet, a sand, and a light combustible material. The "oversize" is passed through a secondary shredder and air classifier. The resulting light fraction is screened. Material retained on the screen becomes a part of the total RDF production. The undersize is returned to the resource recovery module. The heavy fraction from the secondary air classification is processed as is indicated in Figure 11.

*2. Receiving and Firing Facilities*

Despite the fact that as of this writing the power generation plant is in the "shake-down" stage, verbs used in the description and statements that follow are all in the present tense.

A schematic diagram of the RDF receiving and handling facility at the power plant is shown in Figure 12. The four boilers designed for cofiring RDF and coal are rated 42 to 75 MW. The boilers are tangentially fired, suspension firing units, equipped with coal nozzles at each corner. RDF nozzles are installed in two opposing corners. Plans call for the installation of RDF nozzles in the other two corners so that each of the four corners eventually will have an RDF nozzle.

## REFERENCES


1. St. Louis Demonstration Final Report: Refuse Processing Plant Equipment, Facilities and Environmental Evaluations. EPA-600/2-77-155a, December 1977.
2. The journal of resource recovery, *N.C.R.R. Bull.*, 10(3), 1980, 57.
3. **Even, J. C., Adams, S. K., Gheresus, P., Joensen, A. W., Hall, J. L., Fiscus, D. E., and Romine, C. A.,** Ames Evaluation of the Ames Solid Waste Recovery System, EPA Grant No. R803903010, December, 1977.
4. **Trezek, G. J., Diaz, L. F., Savage, G. M. and White, R.,** Prediction of the impact of screening on refuse-derived fuel quality, prepared for Electric Power Research Institute, Report No. EPRI FP-1249, November 1979.
5. **Savage, G. M., Diaz, L. F., and Trezek, G. J.,** RDF: quality must precede quantity, *Waste Age,* 9(4), 100, 1978.
6. Study of processing equipment for resource recovery systems, volume III, Field test evaluation of shredders, prepared by Cal Recovery Systems, Inc., under subcontract to Midwest Research Institute for U.S. EPA, April 1979.
7. **Savage, G. M., Diaz, L. F., and Trezek, G. J.,** Performance characterization of air classifiers in resource recovery processing, Proc. 1980 ASME Natl. Waste Process. Conf., May 1980, 339.
8. **Levy, S. J. and Rigo, H. G.,** Resource Recovery Plant Implementation: Technologies, EPA-SW-157.2, 1976.
9. Conversion of Navy Waste to Densified Refuse-Derived Fuel by the Papakube Process and Identification of Commercial Sources, prepared by Cal Recovery Systems, Inc. for the U.S. Naval Facilities Engineering Command, Contract No. N62583/79 MR543, July 1979.
10. **Trezek, G. J., Savage, G. M., and Jones, D. B.,** Fundamental Considerations for Preparing Densified Refuse Derived Fuel, EPA Grant No. R-805414-010, 1981.
11. **Degler, G. H., Rigo, H. G., and Riley, B. T.,** A Field Test Using Coal: dRDF Blends in Spreader Stoker-Fired Boilers, EPA-600/2-80-095, August 1980.
12. **Hall, J. L., Severns, G. A., Joensen, A. W., Van Meter, D. B., and Shanks, H. R.,** Operation and Emission of a Stoker-Fired Boiler While Burning Refuse Derived Fuel and Coal Mixtures, ASME Paper No. 78-WA/APC-2, ASME Winter Annu. Meet., San Francisco, Calif., December 1978.
13. **Petersdorf, R. J., Sansone, S. M., Plumley, A. L., Rocniak, W. R., and McGowin, C. R.,** Co-firing coal and refuse-derived fuel in a utility steam generator: operational experience and corrosion probe evaluation, paper presented at the Am. Power Conf., April 21 to 23, 190.
14. **Joensen, A. W., Hall, J. L., Even, J. C., Van Meter, D. B., Adams, S. K., Gheresus, P., Severns, G., and White, R. W.,** Evaluation of the Ames Solid Waste Recovery System: Refuse Processing Plant and Suspension Fired Steam Generators, Year III Report, EPA Grant No. R803903-010-0 IERL/EPA, Cincinnati, Ohio, January 1980.
15. **Robertson, C. A. M.,** Emission of HCL from municipal incinerators and related studies, *Solid Wastes Manage.,* 64, 139, 1974.
16. **Savage, G. M. and Trezek, G. J.,** Screening shredded municipal solid waste, *Compost Sci.,* 17(1), 7, 1976.
17. **Hall, J. L., Severns, G. A., Shanks, H. R., Joensen, A. W., Van Meter, D. B., and Olexsey, R.,** Environmental emissions from a suspension fired boiler while burning refuse derived fuel and coal mixtures, Proc. 1980 AMSE Natl. Waste Process. Conf., Washington, D. C., May 11 to 14, 1980, 497.
18. **Fiscus, D. E., Joensen, A. W., Chantland, A. O., and Olexsey, R. A.,** Evaluation of the performance of the disc screens installed at the city of Ames, Iowa resource recovery facility, Proc. 1980 Natl. Waste Process. Conf., Washington, D. C., May 11 to 14, 1980, 485.
19. **Attig, R. C. and Duzy, A. F.,** Coal ash deposition studies and application to boiler design, *Am. Power Conf. Proc.,* 41, 290, 1969.
20. **Reid, W. T.,** Correlation of fuel ash compostion with boiler operations, paper presented at the Electric Power Research Institute Utility Seminar, Fort Lauderdale, Fla., January 9 to 11, 1980.
21. **Gorman, P. G., Shannon, L. J., Shrag, M. P., and Fiscus, D. E.,** St. Louis Demonstration Final Report: Power Plant Equipment, Facilities, and Environmental Evaluations, EPA-600/2-77-155b, U.S. Environmental Protection Agency, Cincinnati, Ohio, December 1977, 408.

Chapter 3

# BIOLOGICAL RESOURCE RECOVERY

## I. INTRODUCTION

### A. Advantages and Disadvantages

The extent to which biological systems can be used in the recovering of resources from municipal waste is limited by the nature of the waste, as well as by restrictions in the form of time requirements that are relatively longer than those characteristic of nonbiological systems. Although at least 50% of municipal wastes are biological in origin, a sizeable portion of this fraction is in the form of paper, a material relatively resistant to biological decomposition. On the other hand, the part of the biological waste stream that is not paper (e.g., food wastes and garden debris) is readily decomposable by bacterial action. Nevertheless, even with the latter fraction, biological processing time is reckoned in days rather than in hours, unless heroic measures are taken.

The resistance of paper to microbial attack stems only partially from the complex structure of its constituent cellulose molecules. In fact, microbial enzymatic systems that can bring about a hydrolysis of cellulose to glucose are of common occurrence in nature. The glucose, in turn, is readily utilized by a wide variety of microorganisms. The difficulty is that the cellulose in wastepaper rarely is in a pure form, and as such, readily vulnerable to bacterial attack. On the contrary, within the limits of the size reduction permitted by the economics of waste treatment, cellulose fibers exist in a particulate form too large to permit a simultaneous access of all of the hydrolytic organisms to all the fibers. Therefore, time must be allowed for a sequential breakdown.

Inaccessibility not only is a function of the particle size of the wastepaper, but it also is a result of the molecular make-up of the paper fibers. In general, the typical wastepaper fiber rarely is completely cellulosic in nature. Instead it is a mixture of cellulose and lignin. The structure of the fiber is such that cellulosic molecules are partially masked or shielded by lignin molecules. The difficulty here is that lignin is extremely resistant to microbial degradation. To complicate matters, the masked cellulosic fraction cannot be completely decomposed until the lignin is removed. The difficulty is further accentuated by the fact that a significant amount of the wastepaper is printed paper, and as such is impregnated with inhibitory or even bactericidal compounds.

Processing newsprint biologically is attended by a double handicap in that the paper not only is contaminated with toxic inkstuffs, but it also is largely lignaceous in composition. In fact, it generally is only incompletely, or at the most very slowly, broken down in biological systems.

A result of the prolonged time element is, of course, a need for relatively large-size equipment or multiplication of pieces of equipment. Since equipment is expensive, an economic penalty is a consequence. If the activity must be carried on in a shelter, then the penalty is amplified by the cost of a proportionately larger shelter. Ranking in importance with increase in equipment requirements and enlargement of shelter is that of expansion of land requirement. This latter penalty or disadvantage need not be significantly serious if the resource recovery operation is carried on at an inexpensive land site. Unfortunately, such an ideal situation seldomly is encountered, and hence, increase in extent of land use must be regarded as a disadvantage of biological resource recovery. On the other hand, it should be pointed out that in almost all cases, the capital and operational costs of biological systems are lower than those of their physical or chemical counterparts.

Yet another disadvantage is the generally low monetary value of the biologically recovered product. However, this particular disadvantage may be rapidly disappearing because of changing conditions. As will be pointed out later, supplies of natural resources which not long ago were regarded as being virtually inexhaustible, or in some cases "infinitely" renewable, are now being found to be finite after all. Not only are they proving to be finite, but more alarmingly the point of complete depletion for many is within the foreseeable future. The economic manifestation of this shift from apparent abundance to imminent scarcity is a transition from low market value to an increasingly higher market value. The consequence is that the crossover in terms of transition from unfavorable to favorable economics is coming closer for biological recovery systems.

The misconception that biological systems are less predictable because of being supposedly more erratic in performance and more susceptible to disruption than are physical-chemical systems is readily dispelled. As with any nonbiological system, a biological system will respond uniformly to inputs previously determined to be within the capacity of the system. Similarly, as with a physical-chemical system, a departure from the proper input level will be accompanied by a response that may or may not be expected. In terms of fastidiousness, or to express it differently, of sensitivity, the average physical-chemical system is more adversely responsive than is the average biological system. The reason is that biological systems generally have a greater inherent "buffer capacity" than do nonbiological systems.

Three biological systems for resource recovery from municipal wastes that could be economically feasible under certain circumstances are composting, anaerobic digestion (biogas or methane production), and hydrolysis of cellulose. Of the three, composting has been in practice the longest and its technology is the most fully developed. Biogas production has come into some prominence in solid waste management in recent years because it constitutes a potential source of energy. Hydrolysis of cellulose is beginning to receive some attention, albeit only minor, as a potential food source and alternatively as an energy source.

## II. BASIC PRINCIPLES

Before undertaking a description and discussion of the three biological recovery methods—composting, methane production, and hydrolysis of cellulose—a brief summary of certain of the basic principles common to the three would be useful. The utility rests on the fact that to realistically evaluate existing and proposed biological systems and practices, it is necessary to be acquainted with the microbiological processes involved.

The principles of interest here are those related to the incidence of the microorganisms needed to accomplish the desired treatment and to the environment to which the microorganisms are exposed. Both aspects have a direct and essential bearing upon the success or failure of an operation. However, since incidence of the desired microorganisms, and to some extent appropriate environmental conditions, differ with each of the three methods, those aspects are reserved for the descriptions of the individual methods. On the other hand, because the principles involved in substrate utilization, an important environmental factor, are common to all biological processes, they are described in this section.

### A. Substrate

The substrate is of particular importance because it is the source of nutrients for the microorganisms in question. Ideally, in a biological recovery system, the waste should be of such a nature that it can serve as the sole substrate. The farther the departure from that ideal, the less practical becomes the system.

The substrate may be considered under two aspects, namely, chemical and physical. Molecular structure and elemental make-up are the items of interest in the chemistry of the substrate. Molecular structure determines the availability, and therefore the utility of a compound in a waste to a particular microorganism. Microorganisms can directly assimilate only the very simple compounds, which are used either as a source of energy or to synthesize new cellular material. Therefore, for a complex molecule to be utilized, it is necessary that it be broken down into its constituent units (e.g., protein to amino acids to ammonia), which then can be assimilated. The breakdown is accomplished through a series of enzymatic reactions, with the enzymes being synthesized by the bacteria. The more complex the molecular structure of a waste component, the more extensive and numerous must be the enzyme systems involved, and the more difficult and time-consuming will be the dissimilative process. It follows that the ability of a given group of microorganisms to utilize a certain type of compound is dependent upon the group's ability to synthesize the necessary enzyme complexes. Inasmuch as the enzyme potential varies greatly in extent from group to group, and to a lesser extent in kind, a variety of microbial species must take part in a biological recovery process. Another result is that certain types of wastes can be processed at a far more rapid rate than can be others.

The recognition that enzymes constitute a key factor has prompted the promotion of certain commercial products as beneficial additives purportedly rich in enzymes. With respect to composting and methane production, the utilization of such additives would be a superfluous gesture.

Ranking in importance with molecular structure is the elemental composition of the molecules, and hence the elemental content of the waste component. This is true because the substrate (waste component) is the source of the elements that will be synthesized into microbial cellular material, and of the compounds that will be oxidized to supply the energy needed in that synthesis. Translated into practical operation, the combined elemental composition of a given mix of wastes is a determining factor with respect to rate and extent of the latter's decomposition, and of the relative amounts of individual waste components required to achieve the balance of elements needed for a speedy and efficient process. In composting, it has a decisive bearing on the agricultural utility of the compost product.

*1. C/N Ratio*

The elements of nutritional value to microbes usually are divided into two broad groups termed appropriately micronutrients, (trace elements) and macronutrients. As indicated by the name, micronutrients are elements which are required only in trace amounts. In fact, if present in macroamounts, they may be inhibitory or even lethal to the microbes. Examples are Co, Zn, Cr, and Mo. On the other hand, macronutrients are those elements which are used in relatively large amounts. Two elements clearly qualify for the latter designation. They are nitrogen and carbon. Phosphorus and calcium are intermediate in demand and hence do not readily fit in either of the two categories. The trace elements are used in enzyme synthesis, to some extent in metabolism, and in intra- and extracellular transport mechanisms. Phosphorus is used in the formation of energy-rich compounds within the cells, and thus is essential to an organism's metabolism. Nitrogen is one of the principal elements as well as a distinguishing feature of the protein molecule. Since a microbe's protoplasm is largely proteinaceous in nature, nitrogen is essential to cell survival and reproduction. Excess nitrogen usually takes on the form of ammonia, which may be present in the free (gaseous) or in the bound state, depending upon pH level and temperature of the composting mass. Thus, ammonium exists in the bound form in an acidic environment, especially a moist one. As the pH goes up, that is to seven and above, the ammonia becomes gaseous and may be lost to the atmosphere. Carbon serves both in cell synthesis and in metabolism. In the former, it is used to build up protoplasm, oils, fats, and carbohydrates. In metabolism, carbon is oxidized to yield energy and $CO_2$. Inasmuch as on the average carbon constitutes about 50% of the cellular matter of the microorganisms as well as approximately 25% of the $CO_2$ molecule, carbon is the element required in greatest abundance in bacterial nutrition.

For best nutrition, the elements must be present in the substrate in the correct ratio of one to the other or others. A very important ratio is that of carbon to nitrogen, the C/N. Ideally, the C/N should range from about 20 to 30/1. The variation in level of the maximum ratio stems from the fact that certain carbonaceous substances are more resistant to microbial attack than are others. Inasmuch as the carbon in the resistant fraction of a waste is only slowly available to the microbes, the need to balance it with nitrogen is delayed, and therefore a higher C/N is permissible. If the C/N is too low, i.e., less than 20/1, nitrogen may be present in excess and subsequently lost in the form of ammonia. In anaerobic digestion, such ammonia could be toxic to the microbial population.

With the exception of nitrogen, all of the elements essential to microbial welfare are present in municipal refuse. The nitrogen content of U.S. municipal waste has declined to a level much lower than that which existed a couple of decades ago. Whereas in the early 1950s the C/N ratio of typical refuse averaged from 25 to 30/1, at present (1980) it ranges from 45 to 55/1.

The problem of a high C/N can be alleviated by removing a portion of the carbonaceous material, or by adding nitrogen. Removal of carbonaceous material in a recycling operation can be accomplished through the reclamation of the paper fraction for other uses. Since paper is the principal carbon source as well as the largest single component of refuse, its removal leaves a residue in which the C/N is within an acceptable range. This residue consists mostly of food preparation wastes and garden debris. If nitrogen must be added to a refuse, a good way to do so is by way of a nitrogen-rich waste. Manures and domestic sludge can serve the purpose. Since the extent of the need for manure in a full-scale operation probably would exceed the supply and would involve importing a waste not of municipal origin, of the two wastes, sewage sludge would be the choice. Moreover, sludge would be an especially appropriate choice since it too is a municipal waste, the disposal of which is a problem. In methane production, the sludge should be either raw (primary) or settled activated sludge, inasmuch as the components of digested sludge are relatively stable, having already been submitted to anaerobic digestion.

The utilization of sewage sludge does not constitute an unalloyed advantage, however. The matter of the heavy metal and pathogen content of the sludge demands attention. However, the heavy metal problem becomes minimal when the sludge comes from sewage that is primarily domestic (residential) in origin. Moreover, the refuse serves as an effective diluent of the limited concentration that may be present in such a sludge.

In the absence of a nitrogen-rich waste, the only potential additive remaining would be a chemical nitrogen fertilizer such as urea or ammonium sulfate. If chemical nitrogen must be added, probably the best method of adding would be in the dissolved form at the highest concentration of nitrogen achievable with the nitrogen compound. The reason for relying upon the liquid form is that a more uniform distribution can be achieved than would be possible with the chemical in its dry form. Moreover, if added as a solid in composting, chances are that a significant amount of the chemical would sift out of the pile during handling. If the addition is intended simply to insure that the product has a C/N suitable for crop production or to bring the NPK (nitrogen, phosphorous, and potassium) to legal specifications, the addition should be delayed until the time of bagging or of incorporation into the soil.

*2. Particle Size*

Ranking in importance with the chemical nutrient make-up of the substrate are its physical characteristics. The following three stand out, namely particle size, structure of the material, and moisture content. Inasmuch as the latter two either have been discussed already or will be discuss later, attention at this point is reserved for particle size.

Particle size reduction exerts an influence on a biological process in two ways: (1) natural barriers to bacterial attack are disrupted; and (2) the ratio of surface area to mass is increased. By "barriers" are meant films, waxy coatings, and other surface protectants that could impede access of the microbes to the components to be decomposed.

Of the two effects, the more important is the increase in the ratio of surface area to mass. It is a well-known truism, as well as being intuitively apparent, that the larger the surface area exposed to bacterial attack or chemical activity, the more rapid will be the biological or chemical reaction, as the case may be. Extended to its logical maximum, the implication would be that the more closely the dimensions of a particle size approach the molecular, the correspondingly more rapid would be the reaction. However, practical considerations as well as other physical characteristics and certain operational procedures combine to establish a minimum particle size far larger than the molecular. With regard to practical limitations, two of the more important are energy consumption and monetary costs. Both increase sharply with reduction in particle size. Because of the generally restricted economics of biological systems (applications), the smallest permissible particle size soon is reached.

Chapter 4

# BIOGAS PRODUCTION

## I. INTRODUCTION

In the frantic search for energy sources, wastes receive considerable attention because they are primarily carbonaceous in chemical composition and, hence, can be used directly as fuel to produce heat energy, or by processing, can be converted into combustible gases and liquids. The use of refuse as a fuel is described in another section, along with physical-chemical methods of processing it to form a fuel. In this section the subject matter is the biological processing of municipal wastes to produce the combustible gas, methane ($CH_4$). In the literature, methane produced often is given the descriptive name "biogas", and the process through which biogas is produced is variously known as "anaerobic digestion", "anaerobic fermentation", or "biogasification". The terms are used interchangeably in this chapter.

The history of the anaerobic digestion of municipal sewage sludge is almost as old as that of sewage treatment. The reason is quite apparent when one remembers that a first step in sewage treatment is to remove the settleable solids from the raw sewage and that something has to be done with the removed solids. Because these solids (raw sludge) are aesthetically obnoxious and hazardous to human health, they are difficult to manage. Consequently, one of the first steps in the evolution of modern sewage treatment was the development of a system for treating the raw sludge. Early in the course of this development, it was recognized that anaerobic digestion could serve as a satisfactory method of treatment. Thus, by the time the energy shortage became a reality, the technology for the anaerobic digestion of sludge was well advanced, although many problems inherent in the process remained to be solved. Consequently, it is not surprising that to those searching for ways to convert municipal solid wastes to a transportable and storable fuel, anaerobic digestion seemed a likely prospect in that its primary gaseous product is methane, a gas that has combustion characteristics not much different from those of combustible gases currently in use.

This historial narrative should not fail to note that anerobic digestion was successfully applied to the treatment of the garbage fraction of municipal refuse in the 1950s and even earlier.[1-3] In one of those early operations, not only were wastes treated, but the resulting methane was used to supply a substantial part of the energy required in the operation of the local municipal waste water treatment plant. In more recent years (1966 to 1972), the utility of anaerobic digestion mainly as a treatment method was investigated in considerable detail at the University of California at Berkeley.[4] The principal objective of the university studies was to evaluate conventional anaerobic digestion as a means of treating the slurry that would be formed if municipal refuse were transported hydraulically. The departure from the earlier studies conducted by others was that both rubbish and garbage were digested. The studies demonstrated that no substance in refuse was inhibitory to the digestion process. Not unexpectedly, it was found that certain components were readily digested, whereas others were quite resistant. Finally, the studies also brought to light a few important problems which are described later in this section. The university researchers came to the conclusion that at least within the next 10 to 15 years (i.e., from 1972), anaerobic digestion would not be economically attractive as a method solely for treating solid wastes.

The prospects for anaerobic digestion in solid waste treatment have been brightened somewhat within the past few years as a result of the world-wide frantic search for energy sources. A logical consequence is that the anaerobic digestion of all solid wastes has become attractive to several workers as a subject for research.[5-8]

## II. PRINCIPLES

### A. Definition

Anaerobic digestion may be defined in the sanitary engineering sense as a waste treatment in which liquid or slurried organic wastes are decomposed biologically under strictly anaerobic conditions with a concomitant production of a gas that is mostly methane and of a relatively stable residue.

### B. Description of the Process

Anaerobic digestion characteristically is a two-stage process in which the first stage blends into the second without tangible interruption. The two steps traditionally are known as the "acid phase" and the "methane-producing phase". In a digester operated on a continuous basis, the two phases are not distinguishable because fresh wastes are added to wastes already in the process of being digested. Thus, material in the acid phase is intermingled with other material in the methane-production phase. Because of the dynamics of continuous cultures and the length of detention periods conventionally involved, the external or integrated manifestation in an on-going continuous digester culture is that of the methane phase.

The acid stage is the first step in the digestion process, and as the name implies, it is characterized by an intensive production of acid. In a batch type culture, progress of the acid phase is manifested by a drop in the pH level of the culture, the extent of which depends upon the latter's buffering capacity. Generally the pH declines to about 6.0. Related to hydrogen ion level but not necessarily manifested in pH level is the volatile acid concentration. In conventional anaerobic digestion, the pH level generally is higher than would be expected on the basis of volatile acid concentration. The reason is the higher buffer capacity (high alkalinity expressed as $CaCO_3$) of a typical digester culture. Short-chain fatty acids of "volatile acids", e.g., formic, proprionic, butyric, valeric, and isovaleric, are the predominant acids produced. The microorganisms responsible for the initial breakdown of the waste solids and the production of the organic acids are indigenous to the wastes and are characterized by a rapid rate of growth.

Acids and the $CO_2$ generated by the acid formers serve as a substrate for methane-producing bacteria (methanogens). Methanogens are strict anaerobes, and therefore atmospheric oxygen ($O_2$) is inhibitory to their growth. They obtain the energy needed for their life processes by anaerobically oxidizing the acids and intermediates synthesized by the acid-forming bacteria. The oxidation is coupled with the reduction of $CO_2$ to methane ($CH_4$). Most species of methane producers are very specific in terms of substrate. For example, *Methanobacterium formicum* can oxidize only hydrogen, carbon dioxide, and formic acid. On the other hand, only butyric, valeric, and caproic acids can be oxidized by *M. suboxydans.* The oxidation of organic acids by the methane bacteria frequently is not complete. For example, acetic and propionic acid may result from the anaerobic oxidation of valeric acid. The acetic and propionic acids serve as nutrients for other methane producers. An important point is that since in most cases the energy-yielding processes of the organisms are anaerobic oxidations in which $CO_2$ acts as the terminal electron acceptor, relatively large amounts of $CO_2$ must be available as a nutrient for the growth of the organisms. Inasmuch as $CO_2$ frequently is not formed from the anaerobic oxidation of the oxidizable substrates, the requirement is met by the

$CO_2$ formed by the acid-forming group. An exception is an energy-yielding reaction by many methane-forming bacteria in which acetic acid is fermented with the formation of the end products $CO_2$ and $CH_4$ as follows:

$$\begin{array}{lcl} CH_3 & \longrightarrow & CH_4 \\ | & & \\ COOH & \longrightarrow & CO_2 \end{array}$$

This latter process is a fermentation and is not anaerobic respiration. An example of anaerobic respiration is the incomplete oxidation of alcohol to acetic acid coupled with the reduction of $CO_2$ and $CH_4$ as carried out by *M. omelianskii:*

$$2CH_3\text{-}CH_2OH + CO_2 \longrightarrow 2\ CH_3\text{-}COOH + CH_4$$

Methane also results from the reduction of $CO_2$ with molecular hydrogen as follows:

$$4\ H_2 + CO_2 \longrightarrow CH_4 + CO_2$$

Since the differences between the cultural requirements of the acid formers and those of the methane producers have an important bearing on the production of $CH_4$ and the accompanying breakdown of the material, they should be emphasized here. The acid formers multiply far more rapidly than do the methanogens. A corollary is that they are much more active. Moreover, as a group the acid formers are less fastidious in their environmental requirements (i.e., more catholic in their nutritional sources) and are more resistant to "shock conditions". Because of these divergencies, and especially so with a continuous system, a careful balance must be maintained between the acid phase and the methane phase. If an imbalance occurs, almost invariably it is favorable to the acid formers and detrimental to the methanogens. If the imbalance is not corrected, acid formation continues until the methanogens are killed and the digester becomes "stuck", i.e., ceases to function. It should be noted that if the imbalance is not corrected, eventually even the acid formers are destroyed.

More specifically the differences between the two groups are as follows:

1. The generally diverse collection of microbial species constituting the acid-forming group includes some that are facultative anaerobes, and as such are tolerant of $O_2$. Consequently the presence of $O_2$ does not seriously interfere with acid production. On the other hand, methanogens are obligate anaerobes, and therefore, their activity is severely inhibited by the presence of $O_2$.
2. The two groups differ in the range of pH within which they can survive or thrive. The pH range for acid producers is quite broad, as low as 4.5 for some and as high as 7.5 for others. A more likely lower level is pH 5.5. On the other hand, the permissible pH range for methanogens is only 6.0 to 7.5. Actually inhibition begins when the pH is lower than 6.5 or higher than 7.0.
3. Generally, acid formers tolerate a much higher acid concentration than do methanogens. This is manifested by the continuation of acid production beyond the concentration at which methanogens are adversely affected. Conventionally, the upper level has been established as being about 1500 to 2000 mg/ℓ as acetic acid. However, by careful conditioning (including suitable buffering) and adaptation, a culture of methanogens can be developed that will thrive at concentrations of as much as 10,000 to 15,000 mg acid per liter of culture.

### C. Environmental Factors

Because the effects of pH level, volatile acid concentration, and oxygen level were discussed in the preceding paragraphs, the present section is devoted to the remaining pertinent environmental factors, namely temperature, nutrient composition and concentration, and residence time.

#### *1. Temperature*

There is considerable dispute as to the advantages of thermophilic (50 to 60°C) over mesophilic (30 to 40°C) digestion. Advantages claimed for the former are a more rapid rate of solids destruction, greater gas yield, and a more readily dewatered residue (sludge). A serious disadvantage inherent in a high-temperature operation is the added expenditure of energy and money needed to maintain the high temperature. Another important disadvantage is the rather long time required to adapt a culture to thermophilic conditions, or as is more likely the case, to develop a culture of thermophiles through enrichment. The reason for taking the time to develop a culture is that it would not be advisable to attempt to start a digester at thermophilic temperatures without a fully developed culture of thermophiles. While it is true that with time and patience a culture can be developed, the time lag is impractically long. Moreover, if a catastrophic event were to wipe out the culture, chances are slim that a digester maintained at a high temperature would be nearby to serve as a source of "seed" culture to bring the "stuck" digester back into operation. On the other hand, since most digesters are operated at mesophilic levels, the odds are in favor of the existence of a neighboring digester geared to those levels.

In summary, it may be stated that taking all things into consideration, it can be concluded that the extent of the advantages to be gained from thermophilic digestion does not necessarily warrant the added expense and energy required to attain and maintain it.

#### *2. Substrate*

Certainly readily decomposable (putrescible) materials digest rapidly. Examples of such materials are vegetable trimmings, garbage, and pure cellulose. Wood, on the other hand, is extremely resistant. Since refuse is a mixture of easily, moderately, and slowly decomposable materials, the rate at which it is digested is determined accordingly.

Other factors of importance are the chemical content of the refuse and its physical state. As with composting and with any other biological system, certain amounts of individual nutrients as well as their ratios, one to the other, are vital to the functioning of the system. In methane fermentation, the carbon-to-nitrogen ratio (C/N) is especially important because it has a strong bearing on the balance between acid formers and methanogens. If the C/N is excessively high, nitrogen begins to be a limiting factor. The acid formers, being rapid growers and more vigorous in substrate utilization, are in a better position to compete for the limited nitrogen, and consequently the growth of the methanogens becomes progressively restricted because of nitrogen deprivation. If the carbonaceous material is relatively resistant to microbial attack, then the permissible C/N is somewhat higher. For example, in the University of California studies,[4] a C/N as high as 50/1 did not become inhibitory when newsprint was the most abundant component. Inasmuch as the C/N of U.S. refuse may be as high as 45 to 55/1, to ensure satisfactory digestion either a nitrogenous material would have to be added or some of the paper in the refuse would have to be removed. On the other hand, if the C/N is too low (less than 15 to 16/1), the danger of ammonia "poisoning" arises. As in composting, the nitrogen not used in building cellular material is converted into ammonia and is released as such. Ammonia is toxic to cultures at very minute levels.

Two important physical factors in the digestion of refuse are particle size and physical state, i.e., whether it is in a solid or a liquid form. To repeat the statement made in the section on biological resource recovery, rate and extent of bacterial attack are functions of the ratio of surface area of the individual particles to their mass. The higher the ratio, the more rapid the attack. This feature is especially important in the digestion of refuse because of the presence of many resistant substances. Reduction of particle size lessens the resistance. If the particle size is too large, the time required to attain a significant degree of decomposition could easily be such as to make the required digestion time unfeasibly lengthy. With anaerobic digestion, the maximum permissible particle size is on the order of 1 to 3 cm with green vegetable matter and less than a millimeter or two with wood.

Inasmuch as the existing technology is based on slurried substrate, the refuse must be suspended in water or in wastewater. The solids content of the slurry (feed) should be adjusted to a level at which the digester contents remain at 8 to 10%, although a higher concentration may be recommended by some authorities.[9] For example, the solids content of the feed can be as much as 16% if 50% of the incoming solids is converted to gas. The solids content could be even higher if the rate of conversion to gas were greater than 50%. A practical corollary is that an increase in rate of conversion brings with it a decrease in required digester volume as well as a reduction in operational requirements.

Required digester volume would be much less if the digester culture itself could be maintained at a high solids concentration. However, if the solids concentration is too great, mixing to the degree needed for satisfactory performance becomes difficult. At present, research is being conducted to find methods through which the upper limits of permissible concentration can be raised without impairing digester performance. A logical sequence to this approach is the work being done on the application of "solid substrate" fermentation to methane production. The concept and even the practice of "solid substrate" fermentation are by no means new. Solid substrate fermentation has long been used in applications other than methane production (e.g., silage fermentation).[10-13] Nevertheless, much remains to be done before the approach becomes a feasible one, especially with municipal waste as the substrate.

### *3. Loading and Residence Time*

Loading and residence time (detention period) are environmental factors because the one determines the quantity of substrate available to microbial attack, while the other limits the time during which the substrate is available to the attack as well as the time permitted for microbial reproduction. Although loading and residence time generally are regarded as operational features pertaining to systems based on continuous cultures, the terms may be appropriately applied to batch cultures. The difference is that the permissible range of amounts involved is much wider with batch loading.

#### ***a. Batch Systems***

With batch systems the maximum loading of substrate is that quantity contained in a slurry that would result in the highest solids concentration not inhibitory to the digestion process. This assumes that the substrate slurry is adequately buffered and contains no toxic substances. Residence time would be the time required to convert the decomposable materials into methane, with suitable allowance being made from terminating the process at the point of diminishing returns.

#### ***b. Continuous Systems***

With a continuous culture or system, the latitude is much less. The lower limit for loading is determined by economic feasibility, and the upper by the efficiency of the culture as modified by economic feasibility. Economic feasibility, especially with the expensive facilities employed in methane production, requires that maximum return be obtained per unit of reactor volume. This implies that the amount of raw material processed per unit of time be the maximum that can be done without inhibiting culture performance. As such, the permissible range of loading is rather narrow. The penalty for overloading in anaerobic digestion is rather immediate and drastic when raw sewage sludge is the substrate. It is less so either with the garbage fraction alone or with garbage and organic rubbish combined.

Generally, symptoms of overloading are a gain in volatile acid concentration, a drop in gas production, and a rise in the percentage of $CO_2$ in the gas. In the digestion of refuse, an immediate result of overloading is a drop in volume of gas produced per unit of volatile solids introduced into the digester as well as in percentage destruction of volatile solids introduced. Persistent overloading eventually leads to increased acid concentration and possibly to a drop in pH level. From these symptoms one can correctly assume that the methanogens are the first of the microorganisms to be adversely affected.

The maximum amount of substrate that can be processed per unit of time is a function of size and activity of the appropriate microbial populations. Consequently, loading capacity can be significantly increased by resorting to certain design and operational features. Perhaps the most useful among the latter is the recycling of the microbial population within the system. The steps involved are explained in the section entitled "Design Components," Volume II, chapter 4.

In anaerobic digestion, loading is expressed in weight units of volatile solids (VS) (dry weight) per unit of digester volume (i.e., culture volume) per day (kg VS/$m^3$ volume or capacity/day). With organic refuse (i.e., garbage plus organic rubbish) the range is from 1.05 to 4.05 or 4.86 kg volatile solids/$m^3$ digester culture/day (0.08 to 0.25 or 0.30 lb VS/$ft^3$ digester volume).

Residence time (detention or retention period) is a statistical term which refers to the amount of time the average waste particle (molecule) is exposed to bacterial attack in the reactor, i.e., the average time it remains in the digester. (With respect to the "hydraulics" of the operation it is the throughput rate.) It is statistical in nature because even with the most sophisticated of designs, in a continuous system a portion of the incoming wastes is short-circuited and hence the residence time for it is shorter than the designated time. On the other hand, another portion of the wastes is detained for a time longer than is indicated.

As with loading, the permissible residence time is a function of economics and of culture efficiency. Obviously, to obtain the greatest use from the reactor, wastes must be processed at the maximum rate consistent with satisfactory cultural performance, i.e., at the shortest permissible retention time. Since retention time or period determines quantity of wastes that can be processed within a given reactor capacity, the shorter the retention period the lesser is the required reactor capacity.

The effect of retention time on culture efficiency is related primarily to the rate of reproduction of the various microbial populations and to their temporal and quantitative capacity to utilize the substrate. The retention time must be sufficiently long to permit all of the constituent populations to reproduce themselves. Under environmentally ideal conditions, the proper retention time is one that is equal to the generation time of the slowest growing of the essential microbial populations, which in anaerobic digestion are the methanogens. If the retention time is shorter than the reproduction time, the culture is diluted at a rate faster than can be compensated by growth of the bacteria, and eventually the bacteria are washed out of the system. Ideally, the time should be one in which the microbial populations are maintained at the upper end of the exponential phase of growth. At longer than necessary retention times, the cultures are in the maximum or stationary phase or even in the decline phase. In these latter phases, rate of activity diminishes, and consequently the efficiency of the culture is lowered.

Retention time may be regarded from two aspects, namely hydraulic detention time and microbial biomass detention time. The two retention times may be maintained independently of each other or they may be combined. If the two are combined, length of hydraulic time is based upon that required to maintain the microbial population, and rate of throughput is made a function of rate of microbial growth. The latter is the usual practice in conventional digestion of raw sewage sludge. In applications in which the two periods are maintained separately, the hydraulic time is a function of the time of exposure of wastes to bacterial action required to accomplish the intended waste treatment, namely to utilize the available nutrients in the waste stream. The more intensive the microbial activity, the shorter is the detention time required. The microbial retention time, as stated before, depends upon the rate of microbial reproduction. Separation of the phases is accomplished by removing settleable solids (hopefully, mostly biomass) from the reactor effluent and returning a certain portion to the reactor. The amount returned in relation to the amount discarded determines the length of the detention period. Establishing a separate detention period for the biomass in the anaerobic digestion of refuse is done with greater difficulty than is the case with raw sewage sludge or other rather finely divided wastes. The difficulty is due to the presence of sizeable amounts of nonbiomass in the form of refractory solids and partially decomposed oversized particles in the reactor effluent. The consequence is that the true retention period for the biomass becomes longer than the nominal value.

The detention periods applied in conventional anaerobic digestion of raw sewage solids range from 15 to 30 days, i.e., hydraulic and biomass detention periods being one and the same. Periods for refuse digestion, as named in the literature, fall within the same range. According to the University of California studies,[4] cellulosic materials (cellulose molecules not masked by lignin) are readily digested within 10 days, whereas lignaceous materials (e.g., wood) are only slightly affected at retention times as long as 30 days.

## D. End Products

The end products of anaerobic digestion are a collection of gases and a residue in the form of slurry. The gases are methane (60 to 70%), $CO_2$ (29 to 39%), and trace amounts of hydrogen, hydrogen sulfide, and other gases. Typically the slurry (digested sludge) is characterized by a dark color, at times a tarry odor, and is readily dewatered. It consists of a supernatant characterized by a high BOD. The latter is not surprising since the supernatant contains the solubilized end products of microbial action on the substrate, as well as colloidal substances and dispersed microbial cells. Usually the settleable particles making up digested sludge from refuse digestion are less uniform in size than those in digested sewage sludge. The settleable solids are a mixture of biomass and undigested solids. The ratio of biomass to undigested solids depends upon the ease with which the substrate can be decomposed by the bacteria as well as upon loading and retention period. Light loading and long detention periods make for higher ratios because more time is available for microbial decomposition of resistant components of the incoming wastes.

The extent of destruction of introduced volatile solids is on the order of 60 to 70%. In the University of California studies,[4] the volatile solids destruction averaged about 60% at a detention time of 30 days (temperature 35°C). Pfeffer[8] noted that a reduction of only 43 to 51% of the volatile solids occurred at a detention period of 15 days and temperature of 35°C, and that extending the period to 30 days brought the destruction to 50 to 54%. From each kilogram of volatile solids introduced, 0.375 to 0.525 $m^3$ of gas (60 to 70% $CH_4$) are produced. Finally, when considering gas yields, it should be remembered that actual yields generally are less than those which would be expected on the basis of organic carbon (volatile solids) content.

## III. TECHNOLOGY

### A. Introduction

Despite the long history of the utilization of anaerobic digestion in sludge stabilization, advances made in the technology of the process have been rather modest, especially with respect to lowering capital and operating costs. Moreover, despite the recent interest in the process as a potential method of converting wastes to a useful source of energy, only a minute amount of effort has gone in the direction of improving the economics of the process. This perhaps is due in large part to the stressing of production rather than to the improvement of the means of production. Thus, within the past few years the lion's share of research effort has been directed towards evaluating the suitability of a variety of wastes and specific crops (herbaceous to arboreal) for conversion into methane. Moreover, a large part of this effort tended to end up in the rediscovery of facts and principles long ago described in the older literature. The consequence is that large, expensive concrete structures (reactors or digesters) continue to be the only proven types for municipal size operations. It is true that a spate of devices have been proposed for small-scale applications (small farms, etc.).[14-16] Some of these have proven to be well suited to their particular applications.

Major advances made over the course of the years in sludge digestion—and even they were made as long as a decade ago—have been the introduction of series or multistage digestion. More recently a contact method has come into vogue that is based on the recirculation of biomass. The basic design components for the units involved in the digestion of refuse are for the most part identical with those for the digestion of sewage sludge. The principal difference perhaps is in the method of mixing the digester contents. Since the design components for anaerobic digestion are conventional, they are described in detail in sanitary engineering textbooks (e.g., References 17 and 18), and the interested reader is referred to them. Consequently, only a brief summary of them suffices for the purposes of this book.

### B. Design Components

The chief design components of the digestion unit are the size of the tank reactor, its gas-collection and sludge-heating system, and method of mixing the culture and recirculating the solids. Gas storage may be incorporated in the tank itself or provided separately. Single or multiple digester units may be operated in parallel or in series. In some operations solids are admitted continuously, and in others on an intermittent basis (once or twice a day). Similarly, the solids may be withdrawn continuously or at infrequent intervals, depending upon the method of withdrawing them. The application or nonapplication of these alternatives in systems leads to a classification of digesters into three types: conventional digesters, high-rate digesters, and two-stage (series) digesters. Diagrammatic sketches of the three types are given in Figure 1.

It is important that air be excluded from gas storage facilities, because a mixture of $CH_4$ and air is explosive at concentrations of methane beginning at 5%. The digester itself may be designed to offer temporary gas storage either by providing the unit with a floating cover, or if the digester has a fixed cover, with a gas collection dome. Similarly, separate gas collectors may be equipped with either a fixed cover or a floating cover.

The contents of a digester tend to stratify unless they are stirred or mixed

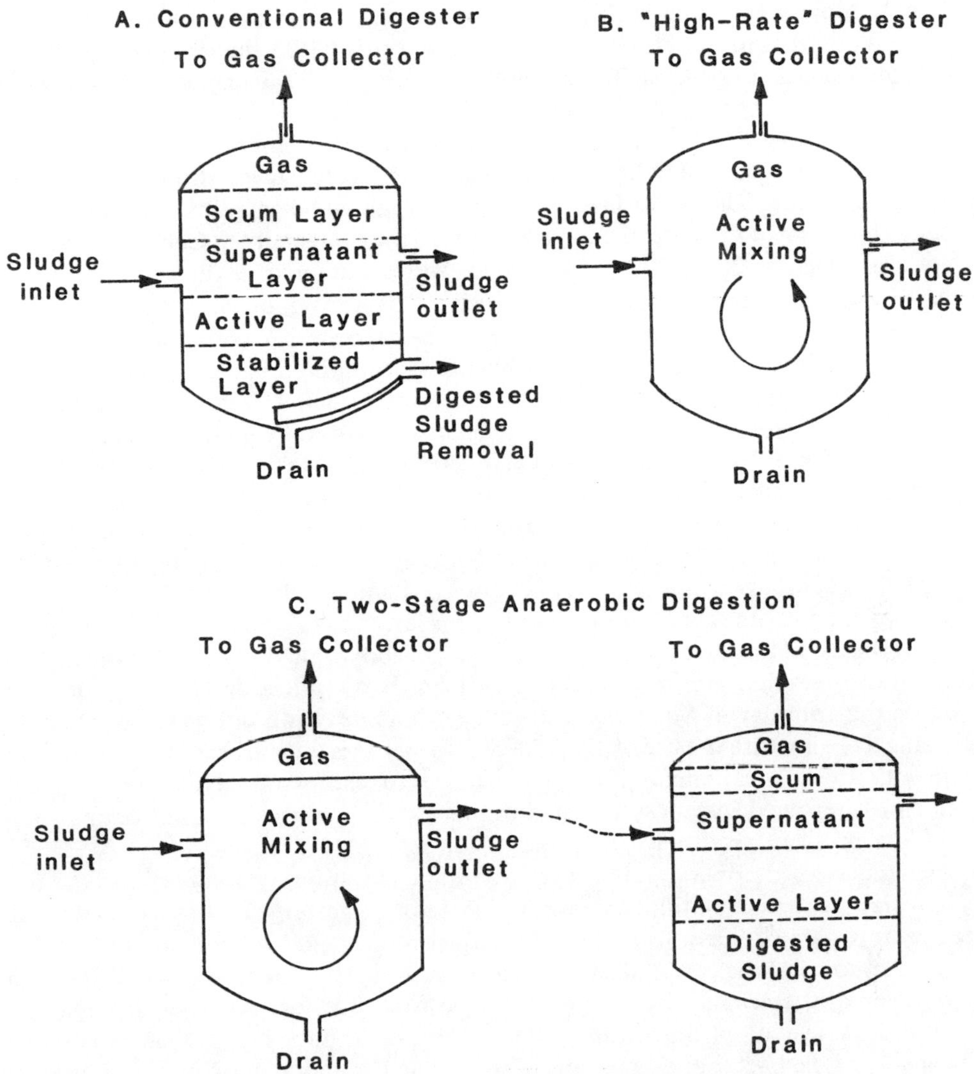


FIGURE 1. Schematic diagram of three types of digesters.

continuously. Thus, in an operation consisting of only a single digester in which the contents are not agitated, a separation into four more or less well-defined strata occurs. As shown in Figure 1, they are (1) an upper, relatively inactive scum layer; (2) a central layer of tank liquor or supernatant; (3) a lower active layer of digesting sludge solids; and (4) a bottom layer of sludge well advanced in stability. With such a digester the incoming solids generally are at middepth. There they are interspersed with the digesting solids and thus are exposed to an active microbial population. Digested solids are withdrawn from the bottom of the unit under hydrostatic pressure. The supernatant is withdrawn through ports located at points between the highest sludge solids and lowest level of the scum layer.

In high-rate digestion, efficiency in terms of volatile solids destruction and gas production is increased significantly by maintaining the culture in a continuously and thoroughly agitated condition. By so doing, active digestion takes place throughout the digester contents, and thus most efficient utilization is made of the digester capacity (volume). Moreover, since such digesters are operated on a continuous basis, an increase is attained in the efficiency of equipment utilization that is characteristic of continuous cultures in general. The agitation disperses both the bacteria and the waste components, and as a result contact between them is ensured. Perhaps the most important consequence of mixing as far as digesting refuse is concerned is the breaking up and reduction of the scum layer. This is important because very little activity takes place in that layer. The scum layer problem is aggravated in refuse digestion by a tendency of the layer to be deep and persistent. In the University of California studies, at times as much as 30 to 40% of the solids were contained in the scum layer.

The physical nature of the scum layer is such that its formation cannot be sufficiently controlled by mixing systems used in conventional sewage sludge digestion. The aggravated tendency to form scum probably arises from the relatively large number of buoyant particles in the slurry. These particles form "nuclei", around which bubbles collect. The persistence probably is a function of the unusually high surface tension of the liquor medium, which contributes to the durability of the individual bubbles that collectively make up the scum layer. If mixing is done by recirculating the digester contents such that slurry is withdrawn from the bottom and reintroduced through the discharge at the top of the culture, the discharge jet of slurry simply bores a channel through the scum layer. The channel acquires a diameter about equal to that of the return stream, and the surrounding scum layer is left intact. Similarly ineffective is the recycling of gas. In the latter case, gas is taken from above and discharged at the bottom of the culture, and then is allowed to move up through the culture and thereby hopefully break up the scum layer. Since the preceding two methods do not work successfully, mechanical mixing with the use of a paddle or a propeller is the only recourse remaining. Even with mechanical mixing, the persistence of the scum layer is not always satisfactorily resolved. As a result, scum formation in the "biogasification" of municipal refuse is a problem awaiting resolution.

When digesters are operated in series (two-stage digestion), the contents of the first in the series are mixed continuously while the second digester serves as a settling unit. The greatest part of the stabilization and conversion to gas takes place in the first digester. In some of the recent designs, provision is made for recycling solids from the active layer of the second digester back to the first digester. Another arrangement involves the recycling of a portion of the sludge to the first digester, and therefore presumably of the bacterial population. This is done through the use of a settling basin in an arrangement comparable to that applied in activated sludge sewage (secondary) treatment.

## IV. SYSTEM PARAMETERS AND CULTURE DEVELOPMENT AND MAINTENANCE

### A. Introduction

The principal system parameters conventionally used in the anaerobic digestion of raw sewage sludge apply equally to that of refuse because the two are based upon identical microbiological principles. The parameters are pH level, volatile solids concentration, buffer capacity (alkalinity), gas production and composition, and quality of the culture suspension and of the effluent. Inasmuch as the principles underlying these parameters have already been discussed, this section is devoted to describing and assigning numerical values to them. An important point to remember in this discussion is that a day-to-day fluctuation is characteristic of all biological systems. However, the fluctuation usually is within a rather narrow range and is such that if an average of the values were taken, the overall trend would be a straight line. Departures from the trend portend trouble and usually are gradual. An abrupt, sharp deviation bespeaks the occurrence of a catastrophic event.

**B. Starting a Digester and Adaptation**

Before embarking upon descriptions of the parameters, a few words on starting a new digester culture are in order. A usually effective method of initiating the development of an anaerobic digestion culture is to introduce into the reactor sufficient digesting culture ("seed") taken from an existing active culture, amounting to at least 10% of the expected volume of the new culture. The "seeding" culture may be obtained from a nearby sewage treatment facility. One of four courses may be followed thereafter.

**Course 1**—Sufficient sewage is immediately added to the seed to bring the liquid volume up to the final volume. (The contents of the digester will then be 10% seed and 90% sewage.) Then loading is begun, but on the basis of the seed content on the first day. Thereafter, it is determined by the amount of the seed plus that of the total volume of feed added previously. This method of gradual increase is followed over a period of time equivalent to that of the designed detention period. Withdrawals are made at the same rates. In other words, the designed detention period is applied immediately but on the basis of seed volumes plus the daily incremental feed. For example, if the volume of the seed used to start culture is 113.5 $m^3$ (30,000 gal) and the designed detention period is 30 days, the volume of feed to be added on day 1 would be 3.8 $m^3$ (1000 gal), and the volatile solids added would be 182 kg (401 lb), based on a loading of 1.6 $kg/m^3$ culture volume per day (0.1 $lb/ft^3$). On day 2, the volume of feed would be based on 113.5 $m^3$ (30,000 gal) plus 3.8 $m^3$ (1000 gal) and therefore would be 117.3 $m^3$ (1033.3 gal). The solids loading would be 184 kg (405 lb) volatile solids. Since the designed culture is 1135.5 $m^3$ (300,000 gal), after day 31 the daily addition of feed would amount to 37.85 $m^3$ (10,000 gal), and the amount of volatile solids 1819 kg (4011 lb).

**Course 2**—Slurried refuse is added to the seed to bring the liquid volume to the final level. No further additions or withdrawals are made until a full complement of microbial organisms has developed. Following the latter event, the digester is ready to be placed in full operation.

**Course 3**—Only the seed is introduced into the digester, after which the feeding schedule described in course 1 is followed. However, no withdrawals are made until the full culture volume is reached. Thereafter the digester is placed in full operation.

**Course 4**—The volume of the seed is that of designed culture volume. Theoretically, the feeding and withdrawal schedule could be initiated on day 1. In practice, however, it would be advisable to allow a brief time for the culture to recover from adverse conditions encountered during the transfer from one digester to the other and to adapt to the new conditions. Moreover, since the microbial populations had been adapted to the utilization of a sewage solids substrate, they would have to be readjusted to the refuse substrate.

Of the four courses, course 1 probably is the best. It provides for a gradual build-up of a microbial population and the adaptation of the population to the refuse substrate. Course 2 is almost as good, but it involves some risk due to the initial abundance of nutrient (refuse), which might make it possible for the acid formers to proliferate at the expense of the slower growing methanogens. Moreover, it is time-consuming in that from 1 to 3 months may elapse before the microbial population reaches the required size. A serious hazard with the third course arises from the initially large volume of free space above the culture. Inasmuch as the air trapped in the space contains the normal complement of atmospheric oxygen, an explosive mixture would result when the culture began to produce methane. The fourth course would be impractical in most cases because of the large volume of seed involved. The problem not only would be in transporting the seed, but also in finding a sufficiently large supply.

The course of the development of a new culture can be monitored by applying the system parameters to be described in the paragraphs which follow. Of the parameters, perhaps the most reliable is the gradual change in gas composition in which the initial predominance of carbon dioxide is superseded by that of methane.

### C. pH

As one would expect on the basis of the microbiology of anaerobic digestion, the pH level of a satisfactorily functioning digester is within the range of 6.5 to 7.5. Because of the greater resistance of the acid formers, the usual symptom of malfunctioning is a continual decline in pH level.

In terms of utility as a parameter, pH level leaves something to be desired, because as a rule a change does not become apparent until any malfunctioning in the culture is well advanced. The explanation of the tardy manifestation is the effective buffering system usually present in a typical digester culture. For example, it is possible to operate a digester at volatile acid concentrations as high as 10,000 mg/ℓ, and yet have the pH of the culture stay within the range of 6.5 to 7.5. Consequently, it is not unusual for a digester culture to become completely stuck and yet have its pH level within a theoretically safe range.

Another explanation for the slow pH response is the fact that pH is expressed as a logarithmic rather than as an arithmetic function. As such it is relatively insensitive, and in digestion, it is affected only by a very large fluctuation in alkalinity.

### D. Volatile Acid Concentration

Volatile acid concentration is a useful parameter because it is a measure of the balance between the activity of the acid formers and those of the methanogens. If the growth of the methanogen population remains synchronized with that of the acid formers, a certain volatile acid concentration is reached, which with a minor fluctuation remains at a relatively constant level. A continued departure from that concentration is a definite indication that one of the two groups, almost invariably the methanogens, is in difficulties. Remedial measures should be taken as soon as the departure is detected.

In sewage treatment practice, a concentration of 2000 mg/ℓ volatile acids has come to be regarded as the upper limit for effective digestion. Concentrations above this level are assumed to be toxic. However, it should be emphasized that high concentrations per se are not necessarily toxic. In an unreported University of California study, gas production and volatile solids destruction continued unabated at acid concentrations of 10,000 mg/ℓ in well-buffered cultures that had been adapted to ultilize the acids. Therefore, rather than being concerned unduly about the degree of concentration, one should rely upon the trend in volatile acid concentration level as an indicator. In fact, if a "steady-state" (i.e., trend) concentration between 200 and 1000 mg/ℓ had become established (the usual range in sludge digestion practice), a digester would be well on the road to becoming stuck by the time the so-called 2000-mg/ℓ danger point had been reached.

The utility of volatile acid concentration as a parameter is somewhat counterbalanced by the difficulty in analyzing for the acids. The test is time-consuming, tedious, and subject to error.

### E. Alkalinity

Alkalinity is an important parameter because it is a measure of the buffering capacity of the culture medium within the neutral pH range, and therefore of most significance to microbial activity. The alkalinity of a digester culture, actually of the suspending medium, is the capacity of that culture to accept protons. In anaerobic digestion, alkalinity results from its bicarbonate, carbonate, and hydroxide components. Of the three, the most important is the bicarbonate because it provides the major share of neutral buffering capacity.

Routinely, alkalinity is determined by titration with a standard solution of a strong mineral acid to the successive bicarbonate and carbonic acid equivalence points, which may be indicated electrometrically or colorimetrically. The analytical procedure is given in detail in *Standard Methods.*[19] Bravko and Chen[20] state that the routine analysis as usually conducted does not give the complete information needed for the safe operation of a digester. They call attention to the fact that as currently practiced, the routine determination of alkalinity by titration to pH 4.0 includes not only all bicarbonate, but approximately 80% of the acetate alkalinity as well. The buffering range of acetate extends only from pH 3.75 to 5.75, a range that is of negligible importance to biological activity. Therefore, they advise that acetate alkalinity ("volatile acid alkalinity") be subtracted from total alkalinity to obtain bicarbonate alkalinity. They state that failure to do so may give the false impression that a digester is adequately buffered. The bicarbonate alkalinity concentration required to maintain a pH of 7.0 is a function of the $CO_2$ content of the digester gas. Thus, at a $CO_2$ concentration of 25%, the required concentration is not quite 2000 mg/ℓ, whereas at about 53%, $CO_2$ it is about 4000 mg/ℓ as $CaCO_3$.

### F. Gas Production and Composition

Gas production and composition are two very useful parameters. Total gas production is a measure of the activity of the entire microbial complex, inasmuch as $CO_2$ is produced both by the acid formers and by the methanogens. Gas composition is important because it distinguishes between methane and $CO_2$. Since methane is a product only of the methanogens, its concentration is indicative of the condition of the methanogen population.

After a "steady state" or trend has been established in rate and amount of gas production and in the composition of the gas, any deviation beyond the upper and lower limits of the normal daily fluctuations within the trend is cause for concern. Generally the first indication of malfunction is an increase in the carbon dioxide concentration at the expense of the methane concentration. This shift may occur before the total gas production drops.

The expected amounts and composition of the gas were given in the section on end products. An appreciable departure from those amounts indicates the existence of inhibitory factors.

### G. Rate and Destruction of Volatile Solids

Inasmuch as the destruction of volatile solids is the result of their utilization by the microbial populations as a source of nutrients, obviously rate of destruction is a function of bacterial activity. Therefore, rate and amount of destruction of volatile solids can serve as a system parameter. Normal rates were given in the section on *end products.*

Directly upon the appearance of the first symptom of culture malfunction, all feeding should be suspended until the adverse symptoms have disappeared. The obvious recourse thereafter is to attempt to discover the cause of the malfunction and then remove the cause.

All too often culture malfunction may be the result of the presence of a bactericidal compound discharged into the sewer, or one that found its way into the refuse. In such a case, about all that can be done is to suspend further feeding and hope that the culture will recover, although concentrations of heavy metals often can be lowered through the addition of sulfides.

Measures taken to raise an undesirably low pH level are only palliative, and as such generally do nothing to remedy an adverse situation. However, if the pH is excessively low, it is necessary to elevate it to a level at which bacterial activity, especially that of the methanogens, may resume. While lime ($Ca[OH]_2$) may be an effective reagent for adjusting the pH, it is beset with a grave potential disadvantage in that the calcium combines with carbonate to form an insoluble precipitate. The precipitate settles to the bottom of the digester. If enough lime is added, the amount of cement-like precipitate formed may be sufficiently great as to interfere with the operation of the digester.

Another method, albeit attended by some risk, is to raise the pH level by adding ammonia. The risk rests in the fact that the $NH_3^+$ ion is very toxic to microbial organisms. Consequently, there is the very real danger of killing off the bacterial populations unless extreme care is taken in adding the ammonia.

Bravko and Chen[20] strongly recommend adjusting an unfavorable pH level through the addition of bicarbonate alkalinity in the form of sodium bicarbonate. According to them, "Bicarbonate is the only chemical in common use that can add alkalinity in any desired quantity and still maintain the physical and chemical integrity of delicately balanced biological systems."

## V. EVALUATION OF THE PROCESS

An evaluation of the potential of anaerobic digestion of municipal refuse rests mainly upon two issues, namely, efficiency of energy recovery and comparative economics. A discussion on these two aspects should in turn take into consideration the fact that anaerobic digestion not only is an energy recovery process, but it also results in a reduction of the waste stream. In other words, anaerobic digestion operations produce a synergy between energy and the environment.[21]

### A. As a Disposal Method

The potential of anaerobic digestion in municipal waste management was thoroughly explored by Klein in the later 1960s.[4] In his investigation, he found that the extent of total and volatile solids destruction of the entire organic fraction of the incoming wastes roughly paralleled that attained in composting, namely from 30 to 60%. If most of the light fraction is removed for other uses (e.g., as refuse-derived fuel, for pyrolysis, etc.), destruction of the remaining organic fraction can be brought up to 70 to 80%.[7] In both cases temperature and detention time have an important impact. Although extent of volatile solids destruction may be comparable to that attained in composting, the costs are much higher. On the other hand, a ready and economically favorable market is available for the methane produced in digestion. Since the solids content of the residue has characteristics similar to those of compost, it can be put to the same uses, although the procedure for application would have to be modeled after that used in spreading sewage sludge on the land. Therefore, whether or not the marketability of the residue would be comparable to that of the compost product remains to be seen. Probably the fairest comparison would be on the basis of residue solids dewatered to a moisture content similar to that of compost.

#### *1. Disadvantages*

The following disadvantages are among those inherent in the application of anaerobic digestion solely to waste treatment: (1) water is polluted; (2) the residue requires treatment before it can be disposed; and (3) the costs involved are not commensurate with the extent of disposal accomplished. The pollution of water is a result of the need to slurry the refuse prior to digestion. This disadvantage can be lessened through the use of sewage as the slurry medium. With such a course, economic feasibility would dictate that the resource recovery plant be in fairly close proximity to a sewage treatment facility. Another possibility is to recycle the slurry medium obtained from dewatering the residue. Of course, a sizeable portion of the liquid is lost with each pass because of the water remaining bound in the slurry. The physical condition of the medium need not deteriorate despite repeated passes, because at each pass it is exposed to the treatment process. A disadvantage is that dewatering adds to the cost of an already expensive process. However, regardless of added cost, dewatering may be essential because of the need to lower transportation costs by reducing the volume of residues, or because conditions at the disposal site necessitate it. The final disadvantage, costs not commensurate with amount of disposal accomplished, becomes especially obvious in a comparison with those of incineration. Volume reduction of combustible wastes through incineration is as much as 90% as compared to 50 to 60% through anaerobic digestion. The total (capital plus operating) cost per metric ton of material processed is $27.50 with the former as compared to $46[22] with the latter. It should be noted that the two cost estimates do not include credits for sale of energy or products, or both. Furthermore, if the Cal Recovery System® approach is used, in which only organic rejects from fiber recovery and RDF production operations are digested, the cost drops down to $25/Mg.[22] In conclusion, it may be stated that the only justification for using anaerobic digestion *solely* as a *treatment* process for refuse would be in conjunction with a hydraulic refuse collection and transportation system.

**As an Energy Recovery Process**

*1. Advantages*

The advantages attributed in the preceding section to anaerobic digestion as a treatment process for refuse also apply to its use as an energy recovery measure. Added to these advantages, such as they may be, is the fact that the energy is recovered in the form of methane. Methane may be combusted directly or, less likely, may be converted to electrical energy in a fuel cell. Yet another use is a feed stock for synthesis of chemicals.

*2. Disadvantages*

In addition to the disadvantages inherent in the utilization of anaerobic digestion as a treatment process are the following two: (1) the useful component, methane, is produced as a part of a mixture of gases saturated with water vapor; and (2) only 30 to 35% of the energy content of the wastes is converted into the energy of methane. In biogas production the ratio of methane to carbon dioxide is partly a function of the composition of the feed material. If the waste is cellulosic in nature, the percentage of methane will be about the same as that of carbon dioxide. If the wastes contain protein or fats, the methane content increases. With the ratio of cellulose to protein and fats found in refuse, the percentage of methane should range from 55 to about 65%. The gas as produced has an energy value of 18,630 to 26,080 kJ/$m^3$ (Btu of 500 to 700 $ft^3$). If the $CO_2$ is removed, the heat value becomes about $38 \times 10^6$J/$m^3$ (1000 Btu/$ft^3$). In addition to methane and $CO_2$, biogas contains trace gases such as $H_2$, $NH_3$, and $H_2S$. The $H_2S$ can lead to considerable corrosion in an internal combustion engine. The $CO_2$ lowers the heat content of the gas. Therefore, to upgrade biogas to pipeline quality, it is necessary to scrub out the $CO_2$ and $H_2S$. Unfortunately, scrubbing involves a significant monetary as well as an energy expenditure (see "Landfill Gas Collection", Volume II, chapter 8.)

The relatively low percentage of energy recovered is probably the gravest disadvantage as far as anaerobic digestion is concerned. Pfeffer estimates that the percentage conversion of the chemical energy of the refuse to that of methane is only on the order of 32.6%.[23,24] An idea of the potential energy to be recovered from a full-scale operation may be gained from calculations made by Pfeffer for a hypothetical [907-Mg/day (1000-ton/day)] plant. He estimates that the total power required to process each ton in such a plant would be 700,283 kJ/Mg (602,045 Btu/ton), assuming 2073 kW (2780 hp) at 30% efficiency. The energy to heat the digester would be about $1.25 \times 10^6$ kJ/Mg ($1.08 \times 10^6$ Btu/ton) at 85% efficiency. Therefore, the total input of energy to process one ton of refuse would be about 1.95 kJ $\times 10^6$/Mg ($1.68 \times 10^6$ Btu/ton). The energy content per ton of refuse *as received* ranges from 0.93 to $1.39 \times 10^4$ kJ/kg (4000 Btu/lb to about 6000 Btu/lb). The energy in the form of methane produced per ton of refuse processed would amount to $2.9 \times 10^6$ to $3.50 \times 10^6$ kJ/Mg ($2.5 \times 10^6$ to $3.00 \times 10^6$ Btu), assuming a methane yield of 83.7 to 112.4 m$^3$/Mg (3000 to 3600 ft$^3$/ton). Therefore, at a maximum yield of methane with the hypothetical plant with the Btu content of the refuse at $9.3 \times 10^3$ kJ/kg (4000 Btu/lb), energy recovery would be on the order of 37.5%. If the energy input for processing is included, the energy recovery would drop to 34%. Energy recovery with the energy content of the refuse at $1.39 \times 10^4$ kJ/kg (6000 Btu/lb) brings the percent recovery down to 25% exclusive of process energy input.

The reasons for the low recovery of energy are twofold: (1) a part of the carbon in the feed is oxidized to carbon dioxide; and (2) a larger part of the energy is locked in refractory compounds and in bacterial cellular material, and thus remains bound in the residue. Pfeffer recommends that the energy in the residue be recovered through incineration designed for steam production. He estimates that the heat energy thus produced would amount to $3.853 \times 10^6$ kJ/Mg refuse. By so doing, percentage energy recovery in his hypothetical plant would be increased to 60 to 70%.

Various schemes have been proposed for increasing the digestibility of the refuse, all of which involve subjecting the waste to physical and chemical conditions conducive to the hydrolysis and partial breakdown of some of the resistant material. The proposed treatments usually involve raising the temperature of the material to 150 to 200°C, applying 125 to 220 psig of pressure, and exposing it to a strong acid and alkali. Two of the leading researchers in this field are Gossett and McCarty.[6] Rather than treat the raw refuse prior to introducing it into the digester, they propose instead that only the residue from the first pass through the digester be subject to the special treatment. The specially treated residue would then be digested. The advantage would be a reduction in the amount of material subjected to the special treatment. Even though the digestibility of the material can be increased and more gas may be produced from treated residue, the gain in gas production most likely would not compensate for the added expenditure. At any rate, the economics are highly doubtful.

*3. Costs*

Estimates of the cost of the energy produced through the anaerobic digestion of refuse as reported in the literature vary widely. However, those ventured in reference[22] seem to be fairly typical. In arriving at the values, the compilers of the report considered a number of options, among which were (1) mesophilic digestion of fluff RDF, with the disposal of the digestion residue into a landfill; (2) mesophilic digestion of fluff RDF, with the subjection of the residue to thermal processing; (3) thermophilic digestion of fluff RDF, with the residue subjected to thermal processing; and (4) mesophilic digestion of Cal Recovery Systems RDF residue, with the fiber fraction being subjected to thermal processing. The RDF here refers to the light fractions from the air classification step. The residue from RDF production in the Cal Recovery Systems® approach is largely organic in nature and contains a minimum of reusable fiber. Because it is a reject, the volumes (or weights) involved are much smaller than those in systems in which fluff RDF is digested. Furthermore, the compilers allowed for credits for sludge disposal, tipping fees, and for the sale of steam.

With the qualifications named in the preceding paragraph, the 1978 cost per million kilojoules with the mesophilic residue-to-landfill option was about $17/million kJ with a 907-Mg/day (1000-ton/day) plant down to $13 with a 2700-Mg/day (3000-ton/day) plant. Assumptions on which were made the estimates for the 907-Mg plant were a total capital cost of $33,400,000, an annual capital recovery cost of $4,131,000, annual operating costs of $7,599,000, cost of RDF of $3,696,000, and annual sludge and tipping fee credits totaling $2,007,000. For the 2700–Mg plant, the assumed costs were a total capital cost of $95 million, and on an annual basis, capital recovery of $11,752,000, operating of $21,586,000, RDF of $6,007,000, and sludge and tipping fee credits of $5,475,000. If the mesophilic residue-to-thermal processing option had been taken, then the 1978 cost per million kilojoules would have been about $15 with a 907-Mg/day (1000-ton/day) plant and $10 with a 2700-Mg/day (3000-ton/day) plant.[22] Total capital costs upon which the estimate for the 907-Mg plant was made was $48,400,000, and for the 2700-Mg plant, $135 million. The breakdown in annual costs for the 907-Mg/day plant was as follows: capital recovery, $6,987,000; operating, $4,298,000; RDF, $3,696,000; and sludge, tipping fee, and steam credits, $3,497,000. For the 2700-Mg plant, it was capital recovery, $16,700,000; operating, $11,304,000; RDF, $6,007,000; and sludge, tipping fee, and steam credits, $10,491,000. With the thermophilic residue-to-thermal processing, the estimated 1978 energy cost dropped to about $13/million kJ with the 907-Mg/day plant to $8.50/million kJ with the 2700-Mg/day plant. Total capital cost upon which the estimate for the 907-Mg plant was made was $48,400,000, and for the 2700-Mg plant, $135 million. The annual costs for the 907-Mg plant were as follows: capital recovery, $6,987,000; operating, $4,298,000; RDF, $3,696,000; and sludge, tipping fee, and steam credits, $66 million. For the 2700-Mg plant, they were the following: capital recovery, $16,700,000; operating, $11,304,000; RDF, $6,007,000; and sludge, tipping fee, and steam credits, $8,619,000. It should be noted that with the mesophilic option, steam credits for the 907- and 2700-Mg plants were respectively $690,000 and $2,070,000, whereas with the thermophilic option, they dropped to $66,000 and $198,000. Estimates made for the mesophilic Cal Recovery Systems® option were based upon a 1978 total capital cost of $22,240,000 for a 907-Mg/day plant and $52,950,000 for a 2700-Mg/day plant.[22] The annual costs for the 907-Mg plant were assumed to be the following: capital recovery, $2,744,000; operating, $2,173,000; RDF, $3,696,000; and sludge, tipping fee, and steam credits, $3,361,000. For the 2700-Mg plant they were the following: capital recovery, $8,550,000; operating, $5,104,000; RDF, $6,007,000; and sludge, tipping fee, and steam credits, $4,084,000.

*4. Compared to Pyrolysis*

The comparison covers three aspects: environmental impact, utility of the end products, and economics. The major impact of anaerobic digestion is the pollution of water. As mentioned earlier, the water in question is that required to suspend the refuse. The required amount varies according to the solids concentration of the slurry. For example, at a 10% concentration, 9 $m^3$ (2396 gal) of water are required per metric ton refuse slurried. If the process can be made to function at concentrations suggested by Pfeffer (e.g., 20%), obviously the water requirement would be lowered accordingly. Reclaiming the water from the residue (sludge) cannot be done cheaply.

The water requirement in pyrolysis is from 1.1 to 2.3 $m^3$/Mg (300 to 600 gal/ton) of refuse processed. Thus, less water is polluted (i.e., required) in pyrolysis.

Air pollution resulting from anaerobic digestion is far less than with pyrolysis, inasmuch as the process is enclosed and gases are collected and stored. If the gas is combusted without prior treatment, a small amount of $SO_x$ and $NO_x$ may be released into the atmosphere. However, if the gas is processed to separate the methane from the other components, the latter end up in the scrubbers (i.e., are converted to water pollutants). No particulates are produced. With pyrolysis, $SO_x$ and $NO_x$ are produced, especially in the pyrolysis-combustion processes. As happens in anaerobic digestion, the gaseous pollutants are converted into water pollutants. However, with pyrolysis and pyrolysis-combustion, particulates are produced that in many cases are difficult to entrain and remove.

Theoretically, and probably actually, the energy return in the form of useful products is greater with pyrolysis. To begin with, in anaerobic digestion only the less resistant fraction (60%) of organic material of biological origin is reacted upon to produce methane, the sole energy source available to the process. It is true that the residue solids can be burned to produce heat energy for steam generation, but incineration is not a part of the digestion process. Indeed, it would be less costly in terms of energy and money to incinerate the nondigestible components prior to digestion. Since in pyrolysis all organic matter, regardless of origin and biological resistance, is converted to energy sources, the percentage conversion may theoretically be as high as 70%. However, it must be mentioned that in favor of anaerobic digestion is the fact that methane has a wider range of uses and is much more adaptable to utilization than are the combustible products resulting from pyrolysis.

The comparison with respect to economics is made quite easily. For a 907-Mg (1000 ton)/day plant, capital and operating costs per metric ton of refuse processed amount to $25 to $50 with anaerobic digestion, as compared to $55 to $66 with pyrolysis or pyrolysis combustion. Allowing for credits in the form of sales of products, etc., the net cost per metric ton, i.e., after sale of the product, is from $15 to $39 with digestion, and with pyrolysis, $26 to $33. The lower cost ($15/Mg) prevails when the Cal Recovery Systems approach is followed.

*C. Status*

Although by its very nature a section on the status of an approach such as anaerobic digestion is out of date almost at the time of its writing, nevertheless it has a historical value, and perhaps even serves as a diagnostic tool. Certainly the information can serve as a base point for evaluating future progress or retrogression in the field.

As stated in the opening paragraphs of this section, anaerobic digestion has a long history of application in the treatment of wastewater sludges. However, the history has not been one of continued or unqualified success. In fact, the search for a more troublefree and less expensive method has been going on for years. Yet, anaerobic digestion continues to be widely practiced, and its continued use now seems to be assured because of one of its products, namely methane.

The prospect of producing a combustible gas, methane, by way of anaerobic fermentation of wastes, is a tantalizing fact in these days of intensive searching for energy sources. As a consequence, several schemes have been proposed for taking advantage of the process. The spread of the schemes range from small installations to handle a ton or less of waste per day to installations large enough to accommodate 100 tons of refuse per day. The greater number of the proposed installations are designed either for manures or for municipal solid waste (domestic refuse). Two of the largest in their respective fields are a plant at Pompano Beach, Florida, designed to process 91 Mg (100 tons) of municipal solid waste per day, and a plant at Bartow, Florida, constructed to handle 23 Mg (25 tons) of cattle waste per day.

The Pompano Beach plant was completed and opened in May of 1978 at a cost of $3.6

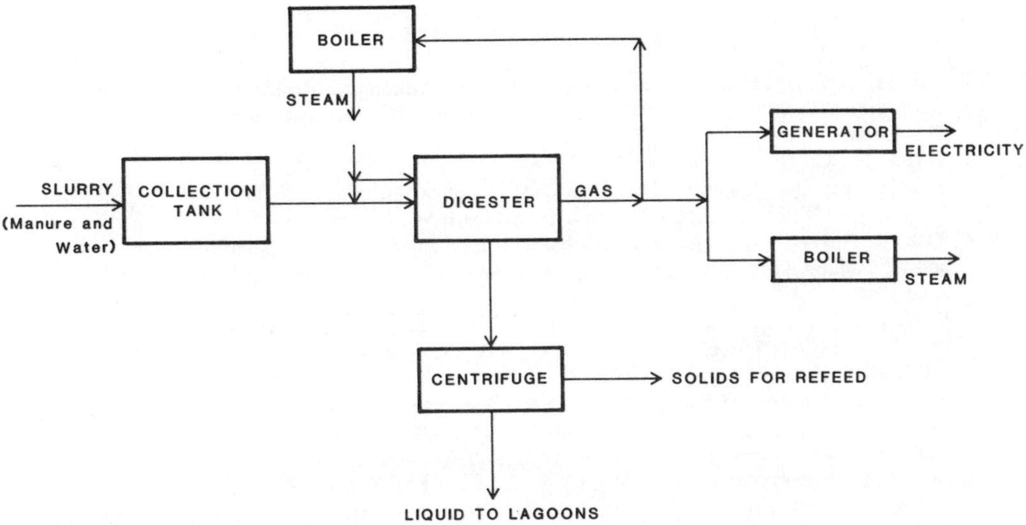


FIGURE 2. Flow plan of anaerobic digestion facility of a slaughterpacking plant feedlot operation.

million. The hope was that with the system, about 85 $m^3$ (3000 $ft^3$) of methane and 85 $m^3$ (3000 $ft^3$) of carbon dioxide can be produced from each ton of refuse processed and that volume reduction of input solids would be about 70%. In the Pompano Beach operation, the refuse is shredded and then is exposed to an electromagnet to remove ferrous metals. The next step is air classification (two stages) to separate the light organic fraction. The latter material is introduced into a premix tank to be blended with sludge, nutrients, and water. The slurried mixture is metered into two mechanically agitated anaerobic digesters. About a 50% conversion of solids to gas is expected. Unfortunately, because of severe operational difficulties due to design deficiencies in the front end of the plant, after a year of operation the plant was as yet far from functioning at the expected capacity.

The Bartow plant was placed in operation in December of 1978. Its cost was on the order of one million dollars. It is a part of the world's largest slaughter and packing plant and it is also the world's largest environmental feedlot. The methane production plant has two 1211-$m^3$ (320,000 gal) tanks, each of which is equipped with a propeller type of mixer. The digestion process is intended to be carried on at a thermophilic level, i.e., as high as 65°C. The digester is designed to be operated at a 10-day detention period. About 23% of the energy produced in the process will be used to produce the steam and electricity required to operate the system. The plan of operation calls for using the digester residue as a feed supplement for the cattle. Whether this latter will prove feasible remains to be demonstrated. Small-scale feeding experiments indicate that only residue from thermophilic digestion can be used as a food (protein mainly) source by cattle and that the nutritional value of the residue for cattle is about 60% that of cottonseed meal. A simplified flow diagram of the anaerobic digestion facility of the operation is shown in Figure 2.

## REFERENCES


1. **Babbitt, H. E., Leland, B. J., and Whitely, F. J., Jr.,** The biological digestion of garbage with sewage slude, *Univ. Ill. Bull.*, 24(24), 1936; University of Illinois Engineering Experiment Station, Bull. No. 287, Urbana, November 1936, 34(24), 3, 1936.
2. **Taylor, H.,** Garbage grinding at Goshen, *Eng. News Rec.*, 127, 441, 1941.
3. **Ross, W. E.,** Dual disposal of garbage and sewage at Richmond, Indiana, *Sewage Ind. Wastes* (presently *J. Water Pollut. Control Fed.*), 26, 140, 1954.
4. **Klein, S.,** Anaerobic digestion, in *Comprehensive Studies of Solid Waste Management,* Golueke, C. G. and McGauhey, P. H., Eds., University of California, 2nd annual report, SERL Report, No. 69-1, June 1969.
5. **Pfeffer, J. T. and Liebman, J. C.,** Biological Conversion of Organic Refuse to Methane, annual report, NSF/RANN/Se/GI-39191/75/2, Dept. C. E., University of Illinois, Urbana report, UILU-ENG-75-2019, September 1975.
6. **Gossett, J. M. and McCarty, P. L.,** Heat Treatment of Refuse for Increasing Anaerobic Biodegradability, progress report of June 1 to December 31, 1974, Technical Report No. 192, Department of Civil Engineering, Stanford University, January 1975.
7. **Diaz, L. F.,** Energy Recovery through Biogasification of Municipal Solid Wastes and Utilization of Thermal Wastes from an Energy-Urban-Agro-Waste Complex, Ph.D. thesis, Department of Mechnical Engineering, University of California, Berkeley, 1976.
8. **Pfeffer, J. T.,** Reclamation of Energy from Organic Refuse, Grant No. EPA-R-800776, April 1973.
9. **Hills, D. J.,** Biogas from dairy and carbonaceous wastes at high solids, Paper No. 79-4582, presented at 1979 Winter Meet. Am. Soc. Agric. Eng., New Orleans, December 14 to 17, 1979.
10. **Hesseltine, C. W.,** Solid-state fermentations, *Biotechnol. Bioeng.*, 14, 517, 1972.
11. **Rhodes, R. A. and Orton, W. L.,** Solid substrate fermentation of feedlot waste combined with feed grains, *Trans. Am. Soc. Agric. Eng.*, 18, 728, 1975.
12. **Hesseltine, C. W.,** Solid-state fermentation. I, *Process Biochem.*, 12, 24, 1977.
13. **Hesseltine, C. W.,** Solid-state fermentation. II, *Process Biochem.*, 12, 29, 1977.
14. **McGarry, M. G. and Stainforth, J., Eds.,** *Compost, Fertilizer, and Biogas Production from Human and Farm Wastes in the People's Republic of China,* International Development Research Center, Ottawa, Can., 1978.
15. **Hayes, T. D., Jewell, W. J., Chandler, J. A., Dell'Orto, S., Fanfoni, K. J., Leuschner, A. P., and Sherman, D. F.,** Methane generation from small scale farms, in *Biogas and Alcohol Fuels Production,* JG Press, Emmaus, Pa., 1980.
16. **Golueke, C. G. and Diaz, L. F.,** Biogasification of wastes, paper presented at the 1981 Biogas and Alcohol Fuels Semin., Chicago, Ill., March 12 to 13, 1981.
17. **Fair, G. M., Geyer, J. C., and Okun, D. A.,** *Elements of Water Supply and Wastewater Disposal,* 2nd ed., John Wiley & Sons, New York, 1971.
18. **Vesilind, P. A.,** *Treatment and Disposal of Wastewater Sludges,* 2nd ed., Ann Arbor Science Publishers, Ann Arbor, Mich., 1979.
19. **Taras, M. J., Greenberg, A. E., Hoak, R. D., and Rand, M. C., Eds.,** *Standard Methods for the Examination of Water and Wastewater,* 13th ed., American Public Health Association, New York, 1971.
20. **Bravko, N. and Chen, K. Y.,** Optimizing gas production, methane content, and buffer capacity in digester operation, *Water Sewage Works,* 124(7), 54, 1977.
21. **Sheaffer, J. R., Ledford, P. A., Nagelvoort, A., Mullan, J. D., and Rozaklis, L. T.,** A Synergistic View of Anaerobic Digestion, report prepared by Shaeffer and Roland Co. for the Corporation of Public Private Ventures, May 1980.
22. Microbial production of methane from refuse, prepared for Southern California Edison Co. by Brown and Caldwell Consulting Engineers, Research & Development Series 78-RD-75, July 1978.
23. **Pfeffer, J.,** Processing organic solids by anaerobic fermentation, in *Proc. Int. Biomass Energy Conf.*, Winnipeg, Can., May 15, 1973.
24. **Pfeffer, J. T. and Liebman, J. C.,** Energy from refuse by bioconversion-fermentation and residual disposal processes, paper presented at the Energy Recovery from Solid Wastes Symp., University of Maryland, College Park, March 13 and 14, 1975.
25. **Diaz, L. F.,** Overview of selected U.S. methane recovery installations, in *Proc. 1979 Biogas Alcohol Sem.*, Chicago, Ill., October 25 and 26, 1979.

Chapter 5

# HYDROLYSIS AND SINGLE-CELL PROTEIN AND ETHANOL PRODUCTION

## I. HYDROLYSIS

Hydrolysis is the first step in biological reclamation processes in which the cellulosic fraction of wastes is converted into useful products. Among the more important of these products are methane, ethanol, and single-cell protein, all of which involve a biological process at some point. The function of hydrolysis is to split the molecule of cellulose, which in its native state is unavailable to microorganisms, into its simpler constituents. The latter then serve as a substrate in the synthesis of the desired product. Hydrolysis is required because the carbon in an intact complex cellulose molecule is inaccessible to the organisms best suited to large-scale methane production, to protein synthesis, and to ethanol production. Since from 50 to 65% of municipal solid wastes are basically cellulosic, hydrolysis becomes an important link in the chain of events which leads to the conversion of that sizeable fraction to useful products.

Since in this section the interest is in single-cell protein and ethanol synthesis, attention is restricted to those aspects of hydrolysis that are pertinent to the two. When considered in this respect, hydrolysis of cellulose usually is referred to as "saccharification" of cellulose.

### A. Cellulose Chemistry

To better understand the mechanisms and the controlling factors involved in the hydrolysis of cellulose, it is necessary to have at least some knowledge of the chemistry of cellulose. It is generally accepted that the cellulose molecule (empirical formula, $[C_6H_{10}O_5]n$) consists of about 3,000 glucose units, corresponding to a molecular weight of roughly 500,000. Cellulose is insoluble in water, alcohol, ether, chloroform, and organic solvents. It is soluble in Schweitzer's reagent, as well as in a mixture of one part of zinc chloride and two parts of hydrochloric acid. It is difficult to hydrolyze. However, long boiling in dilute hydrochloric acid under pressure yields almost the theoretical quantity of D-glucose. The kinetics of hydrolysis point to a b-glucosidic linkage uniformly throughout the cellulose molecule. Exhaustive methylation gives a derivative that corresponds closely to the trimethylated compound, and from which, on hydrolysis, 2,3,6-trimethylglucopyranose is obtained in almost quantitative yield. Consequently, the presumption is that the cellulose molecule consists of a long chain of b-glucose units uniformly linked together in the 1,4-position. Confirmation of the structure of the cellulose molecule has been given by X-ray diffraction patterns. The molecule has a periodicity of repeating unit of 10.3 Å, which is the unit length known to be that of two anhydroglucose units linked in the 1,4-position. A portion of the cellulose molecule is drawn in Figure 1.

Cellulose has two distinguishable regions in its structure, namely crystalline and paracrystalline. Regions of a high degree of order are termed "crystallites" or "micelles". Those in which the cellulose is more randomly oriented are known as "amorphous" or "paracrystalline" regions. The crystalline regions consist almost entirely of true cellulose. Because of the very close packing and the strong forces of attraction by hydrogen bonds and van der Waal forces between cellulose molecules in these regions, crystalline cellulose is more resistant to enzymatic hydrolysis than is the more accessible amorphous cellulose.

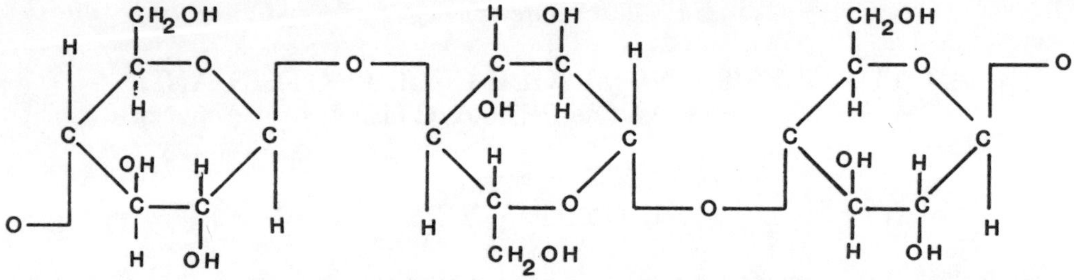

$\beta$-1,4-LINKED GLUCOSE

FIGURE 1. Structure of cellulose molecule.

Other factors related to molecular structure that affect hydrolysis are the surface-to-volume ratio and the presence of associated substances occurring with cellulose in nature. The more important of the latter is lignin. In wood, lignin is present in large amounts and serves to protect the accompanying cellulose from attack by reducing its accessibility.

**B. Types of Hydrolysis**

The degradation of cellulose generally is understood to imply a reduction in the average degree of polymerization (DP) of the polymer. There are four major types of cellulolytic degradation: hydrolytic, oxidative, microbiological, and mechanical. In hydrolytic degradation, the cellulose is reduced in DP and has an increase in reducing power. The cellulose is reduced in DP in oxidative degradation, and the product indicates the development of carbonyl and carboxyl groups. Microbiological degradation is characterized by a reduction in DP, although a loss of strength is the most pronounced effect. In mechanical degradation, the cellulose is reduced in DP by subjecting the fiber to severe physical treatment.

Hydrolysis can be brought about either by physico-chemical (i.e., nonbiological) or by biological means. The latter usually is termed enzymatic hydrolysis. Reagents used in physico-chemical methods may be either acids or alkalis. Each of the two approaches has its advantages and disadvantages. The spreading trend towards enzymatic hydrolysis indicates that for the present there is a strong possibility that the advantages associated with enzymatic hydrolysis outweigh those from physico-chemical hydrolysis.

When comparing chemically-induced hydrolysis of cellulose with that catalyzed enzymatically, differences between the natures of the catalysts must be considered. For example, acid molecules have a much smaller molecular size than that of the proteinaceous enzyme molecule. The molecular weight of hydrochloric acid typically used in acid hydrolysis is 36.5, whereas the molecular weight of the enzyme cellulase is 63,000. Consequently, it can be expected that the penetration by acid molecules into the more resistant regions of cellulose would be greater than by enzymes. However, it has been shown that while hydrolysis may be more complete with the use of acids, particularly if implemented with heat, enzymes are the more efficient hydrolytic agents. For example, at 50°C on a weight basis, $10^5$ times as much acid as enzyme is required to accomplish the same degree of hydrolysis. At the molecular level this difference is even more pronounced in that $10^8$ molecules of acid are required to catalyze the same degree of hydrolysis as one enzyme molecule, and catalysis occurs at a much lower energy level. However, following the reaction, the degree of polymerization of the cellulose molecule is less after acid hydrolysis, probably because the large size of the enzyme molecule limits the latter's ability to penetrate into the cellulose molecule.[1]

### C. Physical-Chemical Hydrolysis

A common method of physical-chemical hydrolysis involves finely milling the waste and immersing it in a weak acid (0.2 to 10% acid) brought up to a moderately elevated temperature (180 to 230°C) and under moderate pressure. The acid hydrolysis of the cellulose can be approximated by a series of first-order reactions summarized as follows:

$$\longrightarrow \text{alphacellulose} \xrightarrow{\text{hydrolysis}} \text{glucose} \xrightarrow{\text{hydrolysis}} \text{mythylhydroxyfurfural} \xrightarrow{\text{hydrolysis}}$$
$$\text{levulinic acid} \xrightarrow{\text{hydrolysis}} \text{formic acid}$$[2]

In practice, the reaction is terminated at the glucose stage.

Lignin that may be with the cellulose does not hydrolyze under the conditions applied in practice.

In his research on the hydrolysis of wood, Saeman[3] followed the series of steps involved in the overall process.

$$(C_6H_{10}O_5)n + H_2O \xrightarrow{\text{acid}} nC_6H_{12}O_6 \tag{1}$$

Alphacellulose constitutes about 40 to 50% of the carbohydrate fraction in wood, and hemicellulose from 20 to 30%. Alphacellulose (true cellulose) is split into its constituent glucose units in hydrolysis. Glucose and pentoses are formed when the hemicellulose is hydrolyzed. The total hydrolysate of wood usually is from 60 to 68% glucose, 2 to 25% mannose, 0 to 6% galactose, 8 to 30% xylose, and 0 to 3% arabinose. Since paper is mostly true cellulose, the primary sugar from its hydrolysis is glucose. From 64 to 75% of the cellulose becomes reducing sugars when hardwoods are hydrolyzed, and from 52 to 70% with softwoods. In terms of fermentable sugars, the respective ranges become 40 to 57% and 40 to 60%. The potential sugar content of ground kraft paper in terms of its cellulose content is 84 to 86% of the weight of the paper, with slurried kraft paper, it is 81 to 85%, and with ground refuse, 38 to 53%. In terms of actual yield, Fagan et al.[4] estimates it to be 52% at 230°C and with 1% acid.

Saeman[3] concluded that the hydrolysis of crystalline wood cellulose can be described by an $A \overset{K_1}{\rightarrow} B \overset{K_2}{\rightarrow} C$ reaction according to Equations 2, 3, and 4:

$$\frac{dC_A}{dt} = K_1C_A \tag{2}$$

$$\frac{dC_B}{dt} = -K_2C_B + K_1C_A \tag{3}$$

where $C_A$ is the cellulose concentration, $C_B$ the concentration of sugar, $K_1$ the reaction rate constant, $K_2$ the reaction rate constant for sugar to decomposition products, and t is time. Equation 4 is

$$K_i = P_i e^{-H_i/RT} \tag{4}$$

where i is 1,2.

Fagan et al.[4] showed that except for a difference in rate constants, the hydrolysis of paper essentially proceeds according to Saeman's model, and that the relationship holds for fibrous samples as well as for ground samples. Through experimentation they determined that the reaction rate can be represented by the following two expressions. The first refers to the decomposition of cellulose (i.e., A → B), and is

$$r_1 = 28 \times 10^{19} C^{1.78} \exp(-45.100/RT) C_A \quad (5)$$

where C is the acid concentration in weight percent, $C_A$ the cellulose concentration weight of residual potential sugar/weight of slurry. The second pertains to decomposition of glucose (B → C), and is

$$r_2 = 4.9 \times 10^{14} C^{0.555} \exp(-32,800/RT) C_B \quad (6)$$

in which $C_B$ is the sugar concentration, weight of sugar/weight of slurry.

In acid hydrolysis, rate of hydrolysis is a function of surface area-to-mass ratio (particle size), liquid-to-solids ratio (L/S), acid concentration, and temperature and time. While the hydrolysis rate advances with increase in the L/S, in practice an upper limit is imposed because of the trade-off between increase in rate and amplification of cost of equipment and operation brought about by the larger volumes to be processed at the higher L/S ratios. The critical point is reached at an L/S of 10/1. Increasing the acid concentration and elevating the temperature brings a corresponding increase in sugar content, but the rate of corrosion also is increased proportionately. The weight of sulfuric acid required for hydrolysis is equal to about 0.5% that of the charge.

Another factor that places a ceiling on acid concentration is the cost of the raw material, especially if the acid is not recovered. Recovery of the acid is difficult to accomplish. Thus far, an acid concentration of 0.5% seems to represent the best achievable compromise between costs and sufficiently rapid hydrolysis. Where energy costs do not contravene, a lower acid concentration and L/S can be compensated by elevating the temperature to about 200°C.

Both the batch and the continuous system are suitable for chemical hydrolysis. The main difference between the two systems is in the equipment. A single reactor is used in the batch system, and all of the steps in hydrolysis take place in the single unit. Much of the continuous system resembles the batch system in that both incorporate similar processing features, namely hydrolysis, flash vaporization, neutralization, and centrifugation. The two systems differ in the design of the reactor. The continuous reactor system consists of a series of individual (three or four) reactor tubes with a screw press after each unit. The hemicellulose portion of charge is hydrolyzed in the first tube, whereas the true cellulose is processed through the second and third reactors. The resulting sugars produced in tube 1 are removed by passing the discharge from the tube through a screw press. The sugars, being soluble, are in the liquid discharge from the press. The solid portion (pulp) is reacidified and is passed through the second and the third reactors. Some systems may incorporate a fourth reactor.

In terms of economics, the batch system is the appropriate choice for operations if the peak capacity is a total of 113 Mg based on two 72.5 Mg/day and using the same storage tanks, and perhaps the centrifuge. The resulting capacity-cost exponent would closely approach 1.0.[5] The amount of sugar produced by a 72-Mg/day plant would range from 2540 to 3266 kg/day.

The production cost in 1968 to 1969 ranged from an estimated \$0.122/kg with a 72-Mg/day plant to \$0.0805/kg with a 907-Mg/day plant. The fraction of the estimated total cost incurred by providing the waste feed ranges from 30% at 72.5 Mg/day to 44.5% at 907 Mg/day. When considering the preceding cost estimates, it should be kept in mind that they are undoubtedly very optimistic, especially when the cellulosic fraction of municipal wastes is the feedstock. For example, the actual cost of attaining the required particle size reduction significantly exceeds that listed in the cost analysis.

**D. Enzymatic Hydrolysis**

As a fermentable carbohydrate, cellulose differs quite drastically from other commonly used carbohydrates in that it is insoluble and is polymerized by a 1-to-4 beta-glucosidic linkage. Besides being characterized by a highly ordered crystalline fraction, it always is associated with hemicelluloses and lignin.[6] In reality, the use of cellulose as a microbial substrate simply adds to the overall mechanism of carbohydrate fermentation in that the cellulose is solubilized for entry into the cellular metabolic pathways. This solubilization is the enzymatic hydrolysis described herein. It is a hydrolysis catalyzed by the cellulase enzyme system of certain bacteria and fungi.

*1. Mechanism*

The many types of microbes that degrade cellulose differ among themselves with respect to their hydrolytic activity, because each of them produces an enzyme system unique in its characteristics. Thus, some microbes produce cellulases that can hydrolyze only soluble derivatives of cellulose, while others synthesize cellulases that are very effective in hydrolyzing cellulose as refractory as native cotton.

On the basis of the two capabilities, a method for categorizing cellulase has been developed. Two separate cellulase enzymes have been proposed,[7] although some researchers[8,9] refer to the two as parts or factors of a complete cellulase enzyme complex. The two parts or systems are the $C_1$ enzyme or factor and the $C_x$ enzyme or factor. The $C_1$ enzyme, or as it is sometimes called the "affinity" factor[9] or "prehydrolysis factor", transforms cotton cellulose and other highly crystalline forms of cellulose into linear anhydroglucose chains. The $C_x$ enzyme (or "hydrolytic factor") then hydrolyzes the linear chains to form soluble carbohydrates, usually cellobiose and glucose. It also acts to hydrolyze reprecipitated or substituted forms of cellulose, such as carboxymethyl cellulose. The relation between $C_1$ enzymatic activity and that of the $C_x$ factor is illustrated by the diagram in Figure 2.

Cellulolytic microbes vary widely as to the total and relative amounts of $C_1$ and $C_x$ activities in their enzyme systems. The initial step which involves the $C_1$ enzyme is slowest and therefore is the rate-determining factor in the hydrolysis of resistant celluloses. Of the many cellulolytic microorganisms tested with respect to their production of $C_1$ cellulase, the fungus *Trichoderma reesei* (formerly *viride*) seems to be the leading contender at present, and hence most of the biological reclamation processes involving saccharification of cellulose are designed around the culture of *Trichoderma*.

The enzymatic hydrolysis of cellulose in a microbial culture generally takes place outside of the cell. With certain bacteria in which the enzymes are confined in the cell wall, hydrolysis takes place only when the cells and cellulose are in close contact. The situation differs with cellulolytic fungi. With them, the cellulose enzymes are extracellular in nature, and therefore the enzyme can be separated from the cells. Obviously, a completely *intra*cellular cellulase cannot be effective in hydrolyzing cellulose because it remains within the microbial cell, and as such it does not come in contact with cellulose external to the cell. Besides being extracellular in nature, fungal cellulases are almost exclusively adaptive rather than constitutive. Since they are adaptive, their synthesis must be induced by a "triggering" agent in the cell's environment.

With respect to *T. reesei,* cellobiose serves both as the chief inducing agent and, under certain circumstances, as an inhibitor of cellulolytic activity. It assumes the latter role when its concentration reaches 0.5 to 1.0%. In fungal growth on cellulose, cellobiose is present as an intermediary at a low concentration level. As a result of the continued growth of the fungi, cellobiose remains present continually at a concentration light enough to act as an inducer, but not heavy enough to inhibit or repress hydrolysis.

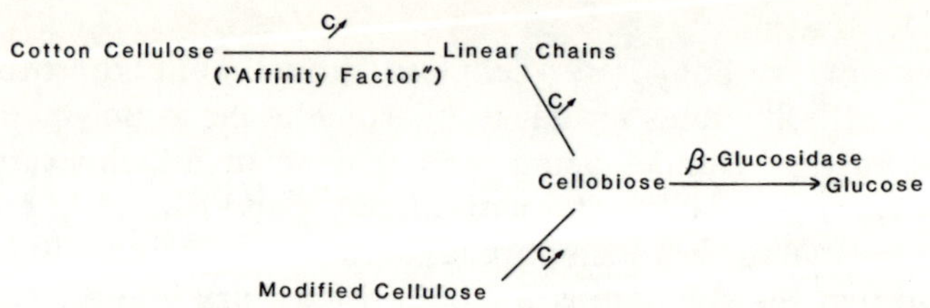


FIGURE 2. Hydrolytic mechanism of cellulolytic enzyme systems.

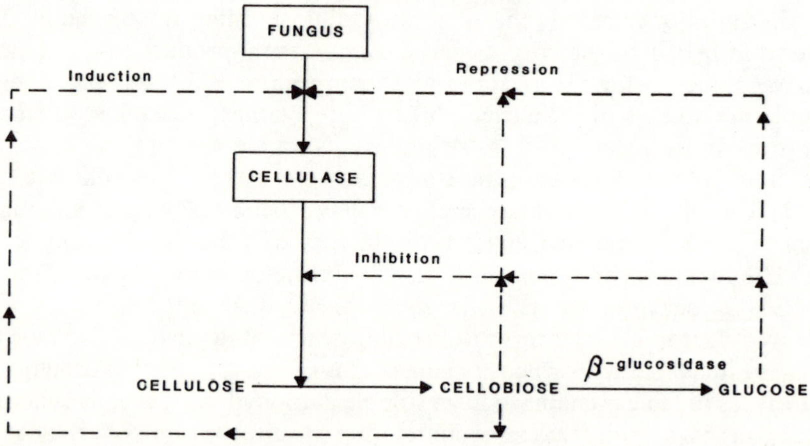


FIGURE 3. Roles of cellobiose in the induction of cellulase in *Trichoderma reesei*.

Simply adding cellobiose to a culture for the purpose of inducing cellulase synthesis would be a futile move because the cellobiose would be very rapidly assimilated by the fungus. At one time it was thought that glucose served as an inducer, but Mandels et al.,[10] in their studies on *T. reesei,* demonstrated that the induction in reality is due to the sophorose present in trace amounts in commercial sugar. A schematic sketch of the roles of cellobiose in the induction of cellulase is presented in Figure 3.

*2. Process Design*

The general process design developed by Rosenbluth and Wilke[11] and by Wilke and Mitra[12] essentially embodies features that could be found in a full-scale operation. Their process has the following features:

1. It provides a two-stage fermentation for enzyme production that involves an initial rapid growth of fungus on glucose in the first stage followed by the introduction of pure cellulose to induce the synthesis of the enzyme in the second stage.
2. The fungus is cultured in a low-cost medium of superphosphate and proflo oil for cell growth.
3. It provides for the recovery and reuse of the original enzyme through precipitation with acetone.
4. Finally, it calls for the combustion of the unhydrolyzed feed material to provide heat needed in the process.

It is summarized by the schematic description presented in Figure 4, and a flow diagram is shown in Figure 5. As the diagram indicates, the process involves several steps which

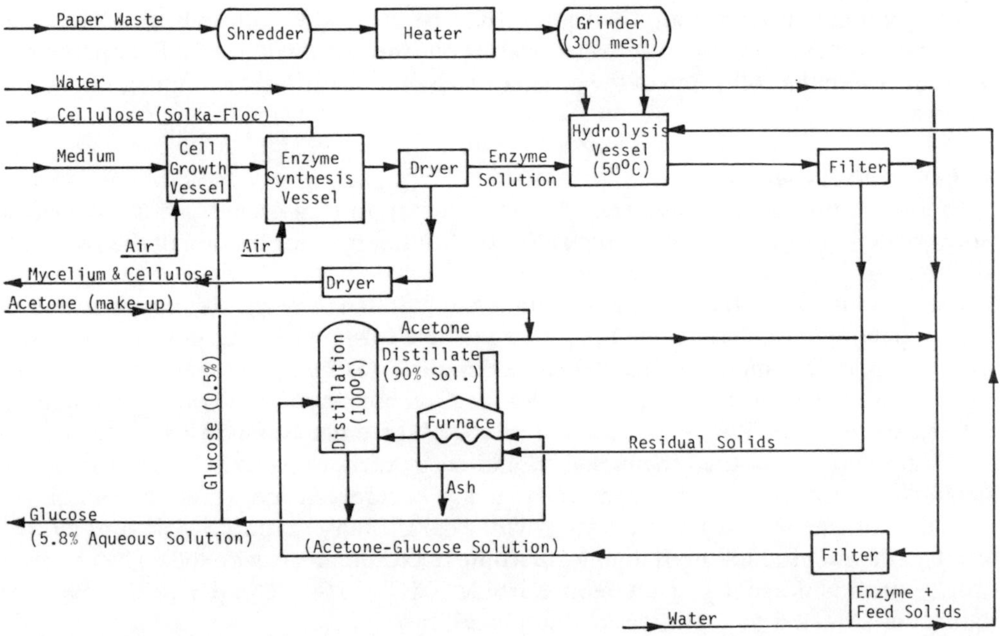


FIGURE 4. Flow diagram of the proposed Wilke-Mitra hydrolysis process.

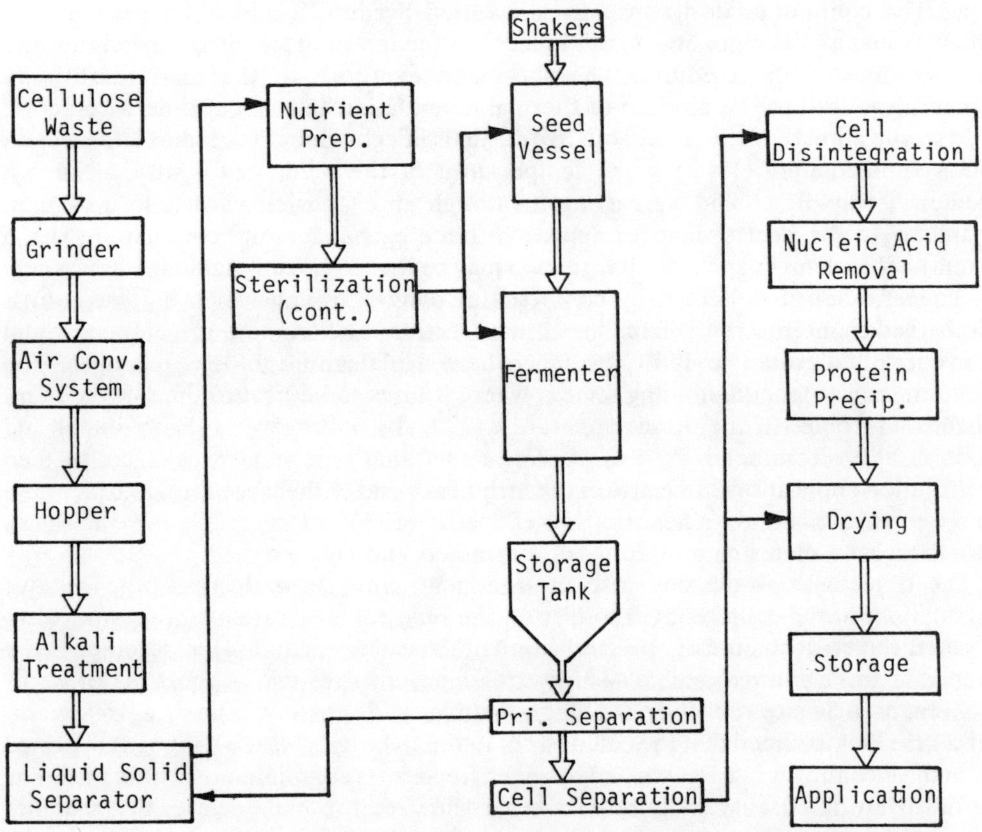


FIGURE 5. Flow diagram of system for the direct production of protein. Modified from Nyiri, L. K., Processing of Cellulosic Materials by Biological Means: An Economic Evaluation of Biosynthesis Systems and Processes, paper presented at the Int. SITRA Symp. Enzymatic Hydrolysis of Cellulose, Hammeenlinna, Finland, March 13 to 14, 1975.

can be grouped into five major divisions. The first major division deals with feedstock preparation; the second with enzyme production; the third with the actual hydrolysis; the fourth with the collection of the glucose product; and the fifth with the drying of the residue.

***a. Feedstock Preparation***

Feedstock preparation involves: (1) the formulation of a growth medium to meet nutritional requirements of the fungus; (2) the continuous sterilization of the medium; and (3) the pretreatment of the cellulose.

The composition of the growth medium recommended by Mandels and Weber[13] is as follows: $KH_2PO_4$, 2 g/ℓ; $(NH_4)_2SO_4$, 4 g/ℓ; urea, 0.3 g/ℓ; $MgSO_4{\cdot}7H_2O$, 0.3 g/ℓ; $CaCl_2$, 0.3 g/ℓ; $FeSO_4{\cdot}7H_2O$, 5 mg/ℓ; $MnSO_4{\cdot}7H_2O$, 1.56 mg/ℓ; $ZnSO_4{\cdot}7H_2O$, 1.4 mg/ℓ; and $CoCl_2$, mg/ℓ. Water serves as the suspending medium. Glucose may be used as a carbon source at the start of a run. While the initial presence of some glucose may bring about a far more rapid growth of the fungus than would occur without glucose, this advantage is counterbalanced by the intervention of a lag period between growth (i.e., after depletion of the glucose) and enzyme synthesis. Protease peptone (Difco®, No. 2) boosts cellulase production. If foaming becomes a problem, it can be solved by adding a silicone antifoam agent (e.g., Union Carbide SAG 471®). The pH of the medium should be adjusted to 5.0 after it is sterilized.

After the medium is made up, it is sterilized either by batch sterilization in holding tanks or by continuous sterilization of either the steam injection or the heat exchanger type. The continuous design usually is selected because it allows for greater heat recovery and makes it possible to minimize the time required for the initial heating and later cooling of the medium. The shortened exposure to the high sterilization temperature lessens the amount of thermal degradation that might otherwise occur.

In sterilization by steam injection, live steam is injected into the medium, thereupon almost immediately elevating the temperature of the medium to 140°C. The hot medium is rapidly cooled by passing it through an expansion valve into a vacuum chamber. In the heat exchanger approach, plate exchangers in combination with a frame heat exchanger and a holding section may be used. The raw medium enters a heat exchanger when it is heated by contact with plates through which hot medium is discharged countercurrently from the holding section. The temperature of the medium is eventually elevated to 140°C by steam heated plates in another exchanger. The medium then enters the holding section where it loses some heat to the raw medium. Finally, it is cooled to operating temperature (29°C) by cooling water. Both Webb[14] and Aiba et al.[15] recommend the use of plate and frame type heat exchangers in feed sterilization applications. Because of the turbulence and of the large actual surface area of the corrugated plates, a heat transfer efficiency of 2511 kJ/m²/hr °C can be attained. Moreover, the plates can be easily disassembled and cleaned.

The objectives of the cellulose pretreatment are (1) to decrease the relative crystallinity of the cellulose; (2) to disrupt the physical structure of the lignin; (3) to lessen the average degree of polymerization of the cellulose; and (4) to accomplish the preceding three at a reasonable cost. Pretreatment involves two or three major steps, namely, particle size reduction (milling), heating, and, possibly, chemical treatment. Of course, it is assumed that the cellulosic material to be hydrolyzed has been separated from the incoming waste stream and has been freed from contaminants within the limits permitted by practicality. With respect to particle size, it should be noted that the finer the particle size, the more rapid is the pace of the process. In addition to increasing the ratio of surface area to contact with fungus or to the catalase synthesized by the organism, milling when accompanied by heating also makes the inner surface of the cellulose more accessible. Moreover, it modifies the structure of the cellulose through the oxidation taking place during the heating. Heating has the additional function of sterilizing the cellulose feedstock. These objectives can be accomplished by maintaining the cellulose at 218°C for 25 min.

An indirect heat calciner dryer is well suited for heating the cellulose, and is to be preferred to a rotary kiln for two reasons. Firstly, the lower gas velocities applied in the calciner dryer are more amenable to use with a light, dusty material such as cellulose. Gas velocities in direct heat dryers must be relatively high because the hot gas is the heating medium. Secondly, the maintenance of sterile conditions is more readily attained when heating is accomplished by indirect heat. The temperature of operation (218°C) precludes the use of a steam-tube heated dryer because of the high steam pressures that would be required.

A pretreatment approach recommended by Callihan and Dunlap[6] is to induce a mild swelling through the introduction of an alkali (e.g., sodium hydroxide). The alkali swells the cellulose and solubilizes part of the lignin. By disrupting the lignin sheathing, the alkali renders the cellulose more accessible to enzymatic action. If the alkaline exposure is followed by air oxidation, the cellulose is degraded to a lower degree of polymerization, and the relative crystallinity is reduced.

***b. Enzyme Production***

The enzyme production step is accomplished in two stages: (1) fungal growth and enzyme synthesis; and (2) separation of the enzyme solution. The fungi are grown in standard industrial fermentor units. Conditions to be maintained in the units are a temperature of 29°C, pH of about 5.0, adequate aeration, and agitation (stirring at 120 rpm). The dilution rate should be on the order of 0.2/day.

A two-stage fermentation system is used. In the first stage, at a dilution rate of 0.2 $hr^{-1}$, the cells grow rapidly on a medium of superphosphate (0.3%), "proflo" oil (0.05%), glucose (0.5%), and the necessary trace elements. Enzyme synthesis is the main feature of the second stage. It is induced through the introduction of pure cellulose.

The enzyme solution can be separated from the cell mass in the suspension by ultrafiltration and by precipitation either with ammonium sulfate ($[NH_4]_2SO_4$) or with acetone. Ultrafiltration, although effective, is prohibitively expensive. Precipitation with $(NH_4)_2SO_4$ is not acceptable because of the difficulty of separating glucose from $(NH_4)_2SO_4$. Thus, precipitation with acetone apparently is the best method at present, even though it too has its troubles. One of the latter is a 14% loss of enzymatic activity with each reuse of the enzyme solution.

The cells and undigested cellulose are removed from the suspension by pressure filter and may be dried for use as a cattle feed.

***c. Hydrolysis***

In the hydrolysis (saccharification) step, the enzyme solution produced in the preceding step comes in contact with the pretreated cellulose waste. The enzymes bring about a hydrolysis of solid cellulose to a final glucose product.

When introduced into the hydrolysis vessels, the ground paper suspension has a solids concentration of about 11.5%. The reaction time is 40 hr. Conditions are maintained aseptic during hydrolysis by the imposition of a temperature of 50°C.

The hydrolyzed effluent is passed through a precoated rotary vacuum filter. In the scheme proposed by Wilke and his colleagues,[11,12] the residual solids are fed to a furnace to provide steam and heat needed for the distillation step. The remaining enzyme solution is fed to the enzyme recovery section. Acetone from the distillation system augmented by a small make-up stream is added to the aqueous enzyme-glucose solution in a volumetric ratio of 2:1. The result is an essentially complete precipitation of protein. The precipitate, which may contain as much as 85% of the original enzyme activity, is recovered through the use of a pressure filter. Enzyme thus recovered is returned to the hydrolysis vessels, and the filtrate is fed to the acetone distillation columns. Heat for the columns and steam for the mycelium dryer can be obtained from combustion of the residual solids. The discharges from the distillation columns are a 90% acetone solution and a glucose solution that contains a negligible amount of acetone. A small portion (1%) of the glucose solution is returned to the first fermentation stage, while the remainder, a 5 to 6% solution of reducing sugar, comprises the final plant product.

## E. Costs

A sizeable element of uncertainty is an almost inevitable accompaniment to the making of cost estimates for processes not as yet applied on a large scale, and this especially holds true with respect to hydrolysis. The uncertainty is a product of errors made in estimating costs. Probably the chief sources of error are a tendency to extrapolate costs of conventional sugar production technology to the saccharification of the cellulose in refuse, and to extrapolate costs based on laboratory-scale results to those of full-scale applications. As a result one may chance upon a reference in which the projected cost of waste cellulose saccharification is stated to be no greater than that of conventional sugar production. An unrealistically low estimate of the total costs for size reducing the waste particles to the dimensions required by the hydrolysis process characteristically is one of the elements in such an erroneous estimate. Another source of error may arise from a failure to make due allowance for the sizeable effort, apart from that for size reduction, required to bring the waste to a condition in which it not only will not adversely affect the biological reactions in hydrolysis, but also the material's handling. For example, toxic (i.e., bactericidal and fungicidal) substances impregnated in or simply soiling the cellulosic wastes can interfere with the enzyme production phase. To procure the essential "cleanliness" is not an inexpensive undertaking. Generally, a safe rule to follow when reading cost estimates in the literature is to keep in mind that the projected estimates are assuredly optimistic ones, and that the actual costs will be much higher.

An idea of the costs involved in the saccharification of waste cellulose, exclusive of procuring the cellulose, may be obtained from the lists compiled by Rosenbluth and Wilke[11] for a facility that would process 9.1 Mg cellulose per day. They are presented in Tables 1 and 2, which are essentially duplicates of tables by Rosenbluth et al.,[11] excepting that English units have been converted into metric units. The facility would produce about 8310 kg glucose (as 5.3% syrup) and 318 kg dry fungal mycelium-cellulose (80% mycelium) per day. This process would involve six main steps:

1. Feed preparation—preparation of nutrient medium
2. Cellulose pretreatment—shredding, heat treated at 200°C, further grinding
3. Microbial growth and enzyme production—*T. reesei* QM6a grown on cellulose along with production of a high level of enzyme induced by the presence of cellulose (residence time, 120 hr)
4. Separation of the enzyme solution—fungal mycelium and residual cellulose filtered from culture liquid containing the enzyme by means of a rotary drum filter, and solid mixture is dried
5. Cellulose hydrolysis (saccharification)—cellulose enzyme solution comes in contact with pretreated cellulosic waste to bring about saccharification to glucose (residence time, 40 hr)

**Table 1**
**MAJOR EQUIPMENT FOR A 9.1-Mg/day SACCHARIFICATION PLANT**

| Item | Size/capacity | Uninstalled cost[a] ($) | Installed cost[a] ($) |
|---|---|---|---|
| Fermenters (4) | 197 $m^3$ | 240,000 | 276,000 |
| Hydrolysis vessel | 263 $m^3$ | 67,500 | 77,600 |
| Filter 1 | 2.23 $m^2$ surface | 21,000 | 27,000 |
| Filter 2 | 8.36 $m^2$ surface | 37,000 | 48,000 |
| Air filter | 0.22 SCMS | 3,000 | 4,000 |
| Shredder | 454 kg/hr | 10,000 | 12,500 |
| Heater | 454 kg/hr | 14,600 | 18,600 |
| Grinder | 817 kg/hr | 25,000 | 31,250 |
| Dryer 1 | 13.9 $m^2$ surface | 12,200 | 14,700 |
| Dryer 2 | 118 $m^2$ surface | 58,000 | 70,000 |
| Heat exchanger 1 | 7.9 $m^2$ | 3,000 | 3,300 |
| Heat exchanger 2 | 4.7 $m^2$ | 2,400 | 2,640 |
| Heat exchanger 3 | 18.6 $m^2$ | 4,500 | 4,950 |
| Heat exchanger 4 | 2.8 $m^2$ | 1,800 | 1,980 |
| Air compressor | 0.234 SCMS | 10,500 | 13,500 |
| Medium supply tanks (2) | 37.9 $m^3$ | 32,400 | 36,900 |
| Fermenter motors (4) | 10 hp | 4,200 | 8,400 |
| Hydrolysis unit motor | 20 hp | 1,500 | 3,000 |
| Medium supply motors (2) | 5 hp | 1,500 | 3,000 |
| Solids feeder | 817 kg/hr | 750 | 850 |
| Screw conveyors (2) | 454 kg/hr | 10,400 | 11,400 |
| Centrifugal pumps (3) | 189 ℓ/min | 2,025 | 2,235 |

[a] 1969 dollars.

Modified from Rosenbluth, R. F. and Wilke, C. R., Comprehensive Stud. Solid Wastes Manage. 3rd Annu. Rep., SERL Report No. 70-2 Sanitary Engineering Research Laboratory, University of California, Berkeley, June 1970, 135.

6. Separation of glucose solution—glucose solution separated from unhydrolyzed cellulose by filtration

## II. SINGLE-CELL PROTEIN PRODUCTION

Studies on the conversion of cellulose and its degradation products to single-cell protein have been concerned mostly with the production of the yeast forms of the fungi, especially *Candida utilis* (otherwise known as *Torula utilis*), and much less so with the production of bacterial proteins (e.g., *Cellulomonas*). Reasons for the preference for the yeast fungi are the demonstrated utility of the product as an animal feedstuff, a rapid growth rate, and a relatively broad tolerance of less than favorable growth conditions. Moreover, certain of the fungi can utilize a broad spectrum of wood-derivative sugars as carbon sources. Disadvantages associated with bacterial production stem mostly from its probable limitations as a feedstuff. The high nucleic content of bacterial cells, the endotoxic content of some, and the possibility of promoting allergenic reactions in humans who handle or consume them render a bacterial feedstuff less desirable than a fungal product. Moreover, bacteria, being much smaller than yeasts, are correspondingly more difficult to harvest.

**Table 2**
**SUMMARY OF FIXED CAPITAL COSTS FOR A 9.1-Mg/day SACCHARIFICATION PLANT**

| Item | Cost ($) | % |
|---|---|---|
| Cellulose pretreatment | 219,000 | 11.1 |
| Shredding | | |
| Heating | | |
| Grinding | | |
| Enzyme production | 1,036,000 | 52.6 |
| Fermentation | | |
| Air and medium sterilization | | |
| Medium supply system | | |
| Cellulose hydrolysis | 245,000 | 12.4 |
| Cellulose recycle and product recovery | 470,000 | 23.9 |
| Filtration | | |
| Drying | | |
| Total | 1,970,000 | 100.0 |

Modified from Rosenbluth, R. F. and Wilke, C. R., Comprehensive Stud. Solid Wastes Manage. 3rd Annu. Rep., SERL Report No. 70-2 Sanitary Engineering Research Laboratory, University of California, Berkeley, June 1970, 135.

### A. Direct Production

In direct protein production, the saccharification step is by-passed and either fungi or bacteria are cultured directly upon nonhydrolyzed cellulose or upon the hydrolysate from cellulose hydrolysis. To be feasible, direct culture requires an organism that is of itself cellulolytic. The yeast *C. utilis* and the bacteria *Cellulomonas flavigena* and *C. uda* meet this condition.

#### *1. Production Systems*

Systems for direct production have been developed and tried on a pilot scale by Nyiri[16] working with researchers at the U.S. Army Natick Laboratories[17] and by Callihan and Dunlap.[6] Actually Nyiri proposed and worked with both the direct production of protein and the "indirect", i.e., culture on cellulose hydrolysate.

Updgraff[18] used *Myrothacium verocaria* as the protein organisms and ball-milled newspaper as the substrate in a laboratory study (i.e., using a 14-ℓ fermenter). It should be noted that newspaper is distinguished by the fact that it has about the same composition as has wood, namely, 40 to 80% cellulose, 20 to 30% lignin, and 10 to 30% hemicellulose and xylosans. The crude protein content of Updgraff's product amounted to about 9%, and the yield, 1.42 g/ℓ. A flow diagram of Nyiri's system is presented in Figure 5.

Although Callihan and Dunlap's system differs in some detail from Nyiri's, the two systems are essentially comparable. Nutrients added to both systems are the same as those used in hydrolysis, and each includes a pretreatment step intended to modify the cellulose structure by reducing its crystallinity. Reduction of crystallinity is done through the use of strongly electrolytic solvents, of which sodium hydroxide (NaOH) is one. The solvent penetrates the cellulose and causes it to swell. Exposure of the cellulose to a 2 to 8% solution followed by air oxidation brings about a swelling of the cellulose followed by a decrease in the degree of polymerization, and a disruption of the physical structure of any lignin that may be in the wastes. The swelling and the disruption of the lignin renders the cellulose more accessible to microbial attack, while the aeration brings about the depolymerization. The effect of the treatment on the cellulose in bagasse is a good example of its effectiveness. Through the treatment, the initial 2% soluble carbohydrate content of bagasse is increased to almost 18%. Moreover, the overall crystallinity is reduced from an original of approximately 50% to a final of 10%.

Callihan and Dunlop used the bacteria *C. flavigena* and *C. uda* in their pilot plant and worked primarily with the sugar cane waste, bagasse. Conclusions reached by the two researchers include the following six:

1. Bagasse must be subjected to alkali treatment before an appreciable bacterial attack can take place.
2. Although 90% of the bagasse may be solubilized, only about 26% is converted to cell mass at a continuous fermenter efficiency of 75%.
3. The usual log-phase doubling time for the culture mass of *Cellulomonas* was from 3.2 to 3.7 hr.
4. Reagent grade salts used in the various experimental media could be replaced by fertilizer grade and industrial grade chemicals.
5. Maximum cell density in pure cultures of *Cellulomonas* was 1.66 g (dry wt)/ℓ, whereas that of a mixed culture (*Cellulomonas* and *Alcaligenes faecalis*) was 6.24 g/ℓ.
6. *Cellulomonas* contains about 50 to 55% crude protein that has a good amino acid balance.

The feed value of Callinhan and Dunlop's product can be gauged from its protein composition (i.e., that of *Cellulomonas* cell protein), which expressed as grams (amino acid per 100 g protein) is arginine 9.21; histidine, 2.30; isoleucine, 4.74; leucine, 11.20; lysine, 6.84; methionine, 1.86; phenylalanine, 4.36; tyrosine, 2.67; threonine, 5.37; and valine, 10.71. The product is a slightly brown to yellow-brown powder.

As Srinivasin[19] states succinctly: "The conversion of cellulosic wastes directly into single-cell proteins is accompanied by many problems." The first problem was mentioned earlier. It is a need for a high degree of cellulolytic activity on the part of the microorganisms employed in the fermentation. Moreover, the organisms must be able to readily and efficiently use the breakdown products as a substrate. A second problem comes from the need to modify conventional submerged culture fermentations to accommodate a solid substrate. A third problem is that the protein of the organism to be cultured must contain a satisfactory amino acid content if the organism is to be used as a source of single-cell protein. It is possible to alter the amino acid composition of the protein to some extent by inducing the organism to produce an intracellular enzyme in large quantities. A fourth and very difficult problem is constituted by the type of protein obtained from the microbial cells. From 40 to 50% of the dry weight of microbial cellular matter is proteinaceous in nature. This by itself would not be a problem except that in addition, 20 to 25% of the dry weight consists of nucleic acids which pose a problem when microbial cells are used directly as animal feed.

*2. Equipment and Costs*

Types of equipment and their relation to the total cost are indicated by the data in Table 3, which are taken from Nyiri's work.[16] The basic flows and inputs for a minimum size commercial plant as proposed by Nyiri[16] are listed in Table 4. (Both tables also include data for "indirect" production, i.e., interposition of hydrolysis.) Nyiri estimates

**Table 3**
**MAIN SYSTEM COMPONENTS IN NYIRI'S SYSTEM FOR YEAST PRODUCTION**

| | Percent of Total Cost | |
|---|---|---|
| | **Direct** | **Indirect** |
| Pretreatment line | 18 | 15 |
| Sterilization system | 3 | 3 |
| Enzyme production line | 0 | 11 |
| Syrup manufacturing line | 0 | 3 |
| Cell (metabolite) line | 22 | 11 |
| Cell recovery line | 5 | 5 |
| Protein recovery | 7 | 6 |
| Drying/distillation | 10 | 10 |
| Instrumentation interface | 15 | 18 |
| Computer hardware—program (software) | 20 | 18 |
| Total | 100 | 100 |

Modified from Nyiri, L. K., Processing of Cellulosic Materials by Biological Means: An Economic Evaluation of Biosynthesis Systems and Processes, paper presented at the Int. SITRA Symp. Enzymatic Hydrolysis of Cellulose, Hammeenlinna, Finland, March 13 to 14, 1975.

that the fixed capital investment for a plant capable of producing 2700 Mg/yr would be approximately $4 million (December 31, 1974 dollars), and the operating cost would amount to $0.24/kg cell mass dry weight.

**B. Indirect Production**

The hydrolysis step described in detail in preceding sections constitutes the first part of the "indirect" production of single-cell protein. Because glucose or another soluble reducing sugar, i.e. the hydrolysate, constitutes the substrate for protein production, a wider variety of candidate microorganisms is a result. However, by reasons of rapidity of growth rate, ease of culture, and feedstuff value of the cell mass product, the choice is fairly well narrowed to yeasts such as *Candida* or *Torula utilis*.

The sequence of steps in the production of single-cell protein on glucose (hydrolysate) is indicated by the diagram in Figure 6. The block labeled "Hydrolysis" represents all of the elements diagrammed in Figures 3 and 4.

Nyiri's estimates[16] serve as an indicator of the costs involved, albeit perhaps overly optimistic. He expects a relative capital cost of a minimum size plant (2700 Mg single-cell protein per year) to be $6,100,000 (1974 dollars), and the direct operating costs, $0.298/kg (dry wt).

## III. ETHANOL PRODUCTION

The interest in ethanol has been intensified of late because of the potential of alcohol as a fossil fuel supplement (e.g., "gasohol"). A natural sequel is a corresponding heightening of interest in the concept of converting cellulosic wastes into ethanol through the fermentation of sugars produced through the hydrolysis of the wastes. As of the time of this writing, the literature on the subject is extremely sparse, and that which

**Table 4**
**BASIC FLOWS AND OUTPUTS FOR A MINIMUM SIZE COMMERCIAL PLANT FOR SINGLE-CELL PROTEIN PRODUCTION**

| | Direct production (Mg) Year Day Hr | "Indirect" (Mg) Year Day Hr |
|---|---|---|
| Raw material input | 8100 27 1.12 | 17,000 57 2.37 |
| Intermediate output | — | 6000 20 832 kg |
| Single cell output | 1200 4 166.4 kg | 1200 4 166.4 kg |
| Liquid flow rate for single cell (l/hour) (average) | 20,000[a] | 20,000[a] |
| Average cell mass (dry weight) | 25 g/ℓ | 25 g/ℓ |

[a] Assuming 20% loss during recovery.

Modified from Nyiri, L. K., Processing of Cellulosic Materials by Biological Means: An Economic Evaluation of Biosynthesis Systems and Processes, paper presented at the Int. SITRA Symp. Enzymatic Hydrolysis of Cellulose, Hammeenlinna, Finland, March 13 to 14, 1975.

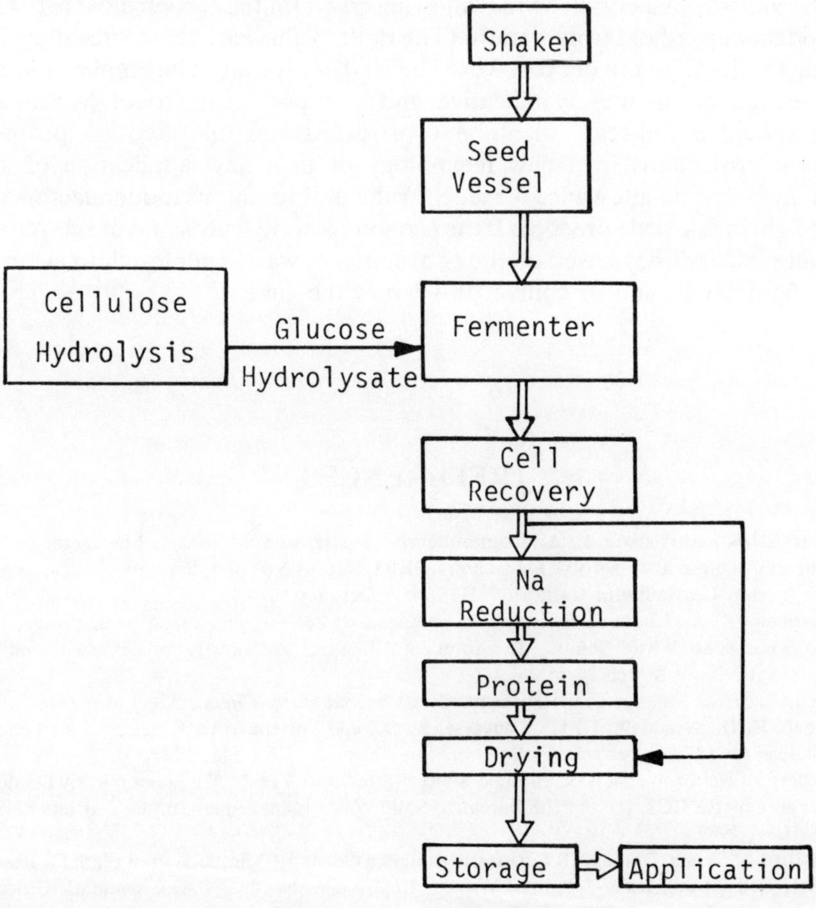


FIGURE 6. Production of single-cell protein on glucose from the hydrolysis of cellulose.

does exist is built largely upon speculation. Of course, the technology of alcohol fermentation is well developed and documented. The scarcity of experience and literature comes in the relation of existing technology to the use of cellulosic wastes, although even in this respect, the uncertainty is more in terms of the preparation of the feedstock for the hydrolysis process than of the utilization of the hydrolysis product (glucose) to produce ethanol. Indeed, there is little reason why conventional fermentation technological principles, procedures, and hardware would not apply here.

The process basically involved is the anaerobic culture of *Saccharomyces cereviseae* on sugars produced in the hydrolysis of cellulose. According to Wilke and Mitra,[12] the hydrolysate sugars are readily fermented by *S. cereviseae* at a theoretical conversion to ethanol efficiency of about 83%. A gross energy conversion efficiency of 47% in the hydrolysis step and of 37% in the entire waste-to-ethanol process is predicted by the two researchers. The net efficiency of the hydrolysis step is 34%, and of the waste-to-ethanol process, 27%. Wilke and Mitra calculated conversion efficiency on the basis of heat of combustion of the feed. The gross values include credit for unused waste heat. The values are an indication of the efficiency that could be reached by adjusting the composition of the wastes such that no unused residual solids remain. The projections are overly optimistic because an unrealistically low energy expenditure is assumed for size reduction, especially when newsprint is involved.

Basing his estimate on glucose being produced at $0.026/kg in the hydrolysis process, and the fermentability of the hydrolysate being 80%, Wilke projected a production cost of $0.11/ℓ anhydrous alcohol without allowing credit for the carbon dioxide ($CO_2$) that is a by-product of alcohol fermentation. (The dollar values are those prevailing in 1973). Allowing for the $CO_2$, the net cost would be $0.08/ℓ. It cannot be emphasized too often that these numbers are merely indicative, and perhaps distantly so, of the actual costs to be met should a full-scale plant be constructed and operated. A problem with theoretical projections in a new technology or in a new application of an "old" technology is that no allowance is made for the multitudinous hidden factors that only come to light in full-scale practice. If the economics were truly so favorable, chances are that a full-scale industry based on the conversion of waste cellulose into alcohol would now be flourishing, and of course, this is not the case.

## REFERENCES


1. **Chan, D. B. and Pearson, E. A.,** Comprehensive Studies of Solid Wastes Management: Hydrolysis Rate of Cellulose in Anaerobic Fermenters, SERL Report No. 70-3, Sanitary Engineering Research Laboratory, University of California, Berkeley, October 1970.
2. **Chapman, R. A.,** Chemical hydrolysis of cellulosic wastes—a kinetic study, paper presented at the Cellulosic Solid Waste Semin., Environmental Control Administration, Public Health Service, Cincinnati, Ohio, March 24 to 25, 1969.
3. **Saeman, J. F.,** Kinetics of wood saccharification, *Ind. Eng. Chem.,* 37(1), 43, 1945.
4. **Fagan, R. D., Grethlein, H. E., Converse, A. O., and Porteous, A.,** Kinetics of acid hydrolysis of cellulose found in paper refuse, *Environ. Sci. Technol.,* 5(6), 545, 1971.
5. **Meller, F. H.,** Conversion of Organic Solid Wastes into Yeast: An Economic Evaluation, report prepared by IONICS, Inc. for the Bureau of Solid Waste Management, under contract PH86-87-204, USHEW, 1969.
6. **Callihan, C. D. and Dunlap, C. E.** Construction of a Chemical-Microbial Pilot Plant for Production of Single-Cell Protein from Cellulosic Wastes, Report SW-24c, USEPA, Cincinnati, Ohio, 1971.
7. **Rosenbluth, R. F. and Wilke, C. R.,** Enzymatic Hydrolysis of Cellulose, SERL Report No. 70-9, Sanitary Engineering Research Laboratory, University of California, Richmond, December 1970.

8. **Leatherwood, J. M.,** Utilization of fibrous wastes sources as sources of nutrients, paper presented at Cellulose Solid Waste Semin., Environmental Control Administration, Public Health Service, Cincinnati, Ohio, March 24 to 25, 1969.
9. **Reese, E. T.,** Mechanistic considerations for the enzymatic hydrolysis of cellulose, paper presented at the Cellulose Solid Waste Semin., Environmental Control Administration, Public Health Service, Cincinnati, Ohio, 1969.
10. **Mandels, M., Parrish, F. W., and Reese, E. T.,** Sophorose as an inducer of cellulase in *Trichoderma viride, J. Bacteriol.,* 83, 400, 1962.
11. **Rosenbluth, R. F. and Wilke, C. R.,** Biological Fractionation of Solid Wastes, in Comprehensive Stud. Solid Wastes Manage., 3rd Annu. Rep., SERL Report No: 70-2 Sanitary Engineering Research Laboratory, University of California, Berkeley, June 1970, 135.
12. **Wilke, C. R. and Mitra, G.,** Process developmental studies on the enzymatic hydrolysis of cellulose paper presented at the Natl. Sci. Found. Spec. Semin., University of California, Berkeley, June 26, 1974.
13. **Mandels, M. and Weber, J.,** The Production of cellulases for specific purposes, paper presented at the Cellulase Symp. Am. Chem. Soc. Fall Meet., Atlantic City, N.J., September 11, 1968.
14. **Webb, F. C.,** *Biochemical Engineering,* D Van Nostrand, London, 1964.
15. **Aiba, S., Humphrey, A. E., and Mills, N. F.,** *Biochemical Engineering,* Academic Press, New York, 1965.
16. **Nyiri, L. K.,** Processing of cellulosic materials by biological means: an economic evaluation of biosynthesis systems and processes, paper presented at the Int. SITRA Symp. Enzymatic Hydrolysis of Cellulose, Hammeenlinna, Finland, March 13 to 14, 1975.
17. **Mandels, M., Lihontz, D., and Nystrom, J.,** Enzymatic hydrolysis of waste cellulose, paper presented at 8th Cellulose Conf., SUNY, Syracuse, N.Y., May 19 to 23, 1975.
18. **Updgraff, D. M.,** Utilization of cellulose from waste paper, *Myrothecium verrucaria, Biotechnol. Bioeng.,* 13, 77, 1971.
19. **Srinivasin, V. R.,** Single cell protein from cellulosic wastes, paper presented at Cellulose Solid Waste Semin., Environmental Control Administration, Cincinnati, Ohio, March 24 to 25, 1969.

Chapter 6

# COMPOSTING

## I. PRINCIPLES

### A. Definition

As one peruses the literature on composting, a variety of definitions of the term "composting" is encountered, some appropriate and others less so. From a solid waste management standpoint, a definition that probably is as suitable as any is the following: "Composting is the biological stabilization of wastes of biological origin under controlled conditions." The key words or phrases in the definition are "biological stabilization", "biological origin", and "under controlled conditions". The term "biological stabilization" limits the process to one in which biological organisms constitute the active agents. "Stabilization" here is interpreted in the sanitary engineering sense, that is, a condition in which the composted waste will not give rise to a nuisance. It does not refer to stability in the chemical sense, i.e., inertness. The term "of biological origin" follows from the condition "biological stabilization". While it is true that a few substances of nonbiological origin can be attacked microbially, generally speaking, only organic materials of biological origin are subject to the biological decomposition needed to attain stability. The specification "under controlled conditions" in an engineering sense distinguishes "composting" from biological decomposition in general. Without the specification, the term "composting" simply becomes a synonym for "decomposition", "decaying", or "rotting". For example, without the specification, an open dump would be eligible for the designation "compost operation". Of course the word "control" is a term of broad meaning, but it does imply the taking of positive measures to promote the decomposition without causing any unduly adverse impact upon the environment. The definition is emphasized because there is a growing tendency to label almost any system involving biological activity as "composting". The reason for the trend is the favorable connotation of the term "composting".

### B. Classification

Composting may be classified on the basis of three broad sets of differences, namely, aerobic vs. anaerobic, mesophilic vs. thermophilic, and windrow vs. mechanical. As the term implies, *aerobic* applies to those systems in which decomposition depends primarily upon the activities of obligate and facultative aerobic bacteria. Conversely, in anaerobic compositing, obligate and facultative anaerobic microbes (acting in the anaerobic phase) carry on the major activity. Because higher temperatures are attained and the nuisance potential is far less, aerobic composting is the preferred method in modern practice. A considerable dispute exists as to the superiority of thermophilic composting over mesophilic composting with respect to rapidity of decomposition. However, it must be admitted that thermophilic composting does have an hygienic advantage in that plant and animal disease-causing organisms, vectors, and weed seeds cannot survive exposure to high temperatures. (It should be emphasized at this point that the mere attainment of high temperatures within a mass of composting material, especially in windrow composting does not guarantee freedom of the entire mass from objectionable organisms.) In a large-scale operation, the matter of which temperature range is preferable becomes immaterial, since in the normal course of events thermophilic temperatures will prevail at about midway through the course of the process. The final division is based upon the extent of mechanization or sophistication of the process. As the term implies, systems involving the use of especially designed mechanized, enclosed units are placed under the heading "mechanical". In windrow composting, after suitable preparation, the material to be composted is stacked in the open in elongated piles (windrows). From the preceding sentences, it becomes apparent that an appropriate synonym for "mechanical composting" would be "enclosed composting", and for "windrow composting", "open composting".

**C. Bacteriology**

Years of experience plus a multitude of scientific studies have shown conclusively that wastes (agricultural and municipal) contain the microorganisms needed to carry on the compost process. In fact, to prevent decomposition, it would be necessary to sterilize the mass. For the purposes of this publication, the reasoning is only briefly stated as follows: As far as the necessary variety of types of microbes is concerned, prior to and at the time of its generation and up to and including its disposal, a waste has had ample opportunity to have been contaminated with the multitude of microbial groups to be encountered in nature. These groups include those which use the waste as a substrate and thereby contribute to the decomposition of the waste. Inasmuch as composting is biological decomposition, the organisms contaminating the wastes have the potential to accomplish the breakdown needed in the compost process. Moreover, since the organisms are indigenous, they are well adapted. A second consideration is that composting is the result of a dynamic succession of organisms in which one group "paves the way" for a succeeding group. As one would expect, there is a considerable overlapping between groups.

An important corollary to the existence of the necessary variety and numbers of indigenous microorganisms is the absence of any need to introduce extraneous organisms, that is to resort to inoculums. Moreover, any effect of the activity of the small number of microorganisms added as an inoculum is overwhelmed by that of the vastly greater number of indigenous organisms. The reality is that the indigenous populations being adapted to the particular environment are in a position to easily compete with the added organisms to the detriment of the latter. Finally, practical experience has demonstrated the superfluity of inocula in composting.

**D. Environmental Factors**

Aside from economics and especially in windrow composting, structure, moisture content, and availability of oxygen (i.e. as $O_2$) all must be considered together when determining the dimensions of a suitable particle size for composting. The reason is that they are interdependent. Perhaps the best point at which to begin to explain the interdependence is the oyxgen requirement. Since modern composting is intentionally aerobic in nature, oxygen must be available to the microbes. In a windrow, the oxygen is obtained from the air entrapped within the interstices, that is, voids between the particles that constitute the pile. The integrity of these interstices depends upon the structural strength of the particles. If the particles are structurally weak, they tend to collapse under pressure into a compact mass. When certain materials (e.g., paper) are wettened, this tendency is aggravated. A severe diminution of interstitial volume results in a correspondingly sharp loss of available oxygen, and the pile becomes anaerobic. A comparable combination of phenomena occurs in mechanical composting, but the expression differs. With mechanical composting, poor structural strength combined with "excess" moisture and the tumbling action imparted by the compost equipment result in the formation of large clods of material, the interiors of which are anaerobic. Translated into general practical terms, the following particle sizes are appropriate for most operations: If the waste contains a large fraction of paper, as is the case with unsorted municipal refuse, then a particle size of 3.8 to 5.0 cm is indicated. On the other hand, if the predominant material is structurally firm, then particle sizes within the range of 0.5 to 1.0 cm are appropriate (i.e., if economics also permit). Especial caution should be taken when dealing with wastes that are largely vegetable in origin. If reduced to a small particle size, a soupy mass results with which it is difficult to contend. (It should be noted that such material should not be size reduced with the use of a hammermill, unless the material is a lesser ingredient of a waste consisting largely of firm components. A machine akin to a silage chopper would be more appropriate.)

*1. Moisture Content*

Theoretically, a moisture content approaching 100% should be optimum. However, in practice the suitable range is much lower and is dependent upon the physical characteristics of the waste. In windrow composting the upper permissible limit is a function of the interstitial volume, and specifically of the extent to which the integrity of the collection of interstices is maintained. As explained in the preceding paragraph, interstitial volume and integrity in turn are functions of structural strength of the particles and of particle size. The limitation placed on moisture, apart from its effect on structural strength of the particles, is that water shares the interstitial volume with air. Thus, the greater the moisture content, the larger is the interstitial volume occupied by water, and correspondingly, less volume remains for air. In terms of actual numbers, the upper permissible range is 55 to 60% when composting municipal refuse, since paper constitutes the greater part of the refuse. In specialized compost operations in which either wood, straw, rice hulls, dry leaves, or another comparable material is the principal constituent, the permissible moisture content may be as much as 85%, although 70 to 80% are safer levels.

By coincidence, the upper permissible moisture content is about the same with mechanized composting. Here, moisture content becomes limiting when it changes the physical nature of the solids such that they form the clods or globs described earlier.

If the moisture content is too low, biological activity is inhibited both in windrow and in mechanical composting. Activity begins to decline gradually at levels between 40 and 50% moisture content, and more rapidly thereafter. However, results obtained in a recent study by the authors, in which refuse "lights" served as a bulking material for composting sludge, indicate that with such a combination, a moisture content as low as 40% would be quite satisfactory. Nevertheless, safety dictates that as a general rule the range of 40 to 50% should be regarded as being the minimum. Biological activity ceases when the moisture content drops to 12% or lower.

When moisture is in excess, it is necessary to take remedial measures. If land and time permit, spreading accompanied by agitation to promote evaporation (drying) would be one recourse. Another method is to add a "bulking" or "absorbent" material. The bulking material not only absorbs the moisture, it also increases the interstitial volume. Examples of bulking materials are sawdust, ground bark, straw, leaves, and rice hulls. If allowed to dry somewhat, the compost product itself makes a satisfactory bulking agent. With municipal refuse, a bulking additive is not needed inasmuch as paper serves the purpose, albeit not as well as a structurally strong material such as sawdust. If a bulking agent should be required, as for example to accommodate a larger ratio of sewage sludge, then the use of the compost product itself is recommended.

*2. Oxygen*

As stated before, in windrow composting the source of oxygen is the air contained in the interstices. Replenishing the oxygen is accomplished by an operational step commonly termed "turning", which is described in a later section. Another method, also to be described later, involves forced aeration. In mechanized composting, oxygenation (aeration) is accomplished by tumbling the composting material, and in certain processes, partly by forced aeration.

*3. Hydrogen Ion (pH)*

Nominally, in the composting of municipal refuse, pH level does not pose a problem. During the first few days after composting is initiated, the pH level usually drops to within the acid range (5.5 to 6.0) as a result of the activity of acid-forming microbes. Despite the drop, the composting process apparently is not impeded. Furthermore, acid formation itself is a form of decomposition. Soon the acids are used as substrate by other microbial forms and the pH rises accordingly. By the time the process is finished, the pH level will have reached 8.5 to 9.0.

Often, lime or some other buffering agent is ill-advisedly added to prevent the initial drop in pH. The qualification "ill-advisedly" is a valid one for two reasons. First, the step is unnecessary, and therefore if done, makes the process more expensive than need be. Secondly, the addition may promote loss of nitrogen by facilitating volatilization of ammonia. The resulting elevated pH, the high temperatures characteristic of composting, and the existence of a C/N lower than 20/1, all combine to bring about the volatilization.

## II. TECHNOLOGY

### A. Introduction

The technology for composting logically is aimed at providing optimum environmental conditions under which the process can proceed efficiently and without adversely affecting the environment or endangering the public well-being. To be practical, the technology must accomplish this function within the *limits* of *economic feasibility*. Inasmuch as the compost product has a low monetary value, and as a waste treatment process composting must compete with other processes, expenditures on the former's technology must be relatively modest.

In composting, the major environmental factor that can be most influenced by technology, and de facto, around which system designs are developed, is the availability of oxygen. The centralized position held by oxygen supply stems from the fact that rapidity of process, elevation of temperature with its beneficial effect on hygienic aspects, and absence of objectionable odors are all dependent upon effective aeration. It is true that moisture has an important role, but that particular requirement can be met with the use of simply designed appurtenances.

Inasmuch as it would be rash to state that any technology is fully developed, the claim made herein is that within the realm of economic feasibilty, the technology for composting has progressed far enough to leave little room for improvement. Any further sophistication would entail monetary expenditures, the size of which would render economic feasibility almost impossible to attain. With respect to design, the equipment for providing aeration ranges from the relatively simple to the very complex. This range leads to the generalized classification of compost technology described earlier, namely "windrow" ("open") and "mechanized" ("closed"). In terms of practicality the important point is how the two classes compare with respect to effectiveness in accomplishing aeration, and even more importantly, how they meet the objectives named in the preceding paragraph. If no significant improvement of the process and its product over that with a windrow system accompanies the use of a mechanized system, then the windrow system should be adopted, because the economics of the mechanized approach are less favorable.

The two major preparatory steps in a compost operation, windrow or mechanized, were alluded to in the discussion on substrates. They are sorting to remove noncompostable and salvageable components, and size reduction. The technology involved in accomplishing these steps is described in the section on size reduction and air classification.

## B. Windrow

### *1. Site Preparation*

Site preparation for a windrow operation is a relatively simple step. A hard surface (e.g., asphalt capable of supporting movable equipment during rainy weather) should be provided. The surface area required is one large enough to accommodate the composting material during the more active period in composting, that is during the time in which equipment must have access to the windrows. An approximation is about 4 ha/100,000 population served. Although theoretically no drainage or seepage should come from the piles, arrangements should be made for intercepting and collecting seepage if it should occur. Ideally, but not necessarily, the surfaced area should be sheltered by a roof. Roofing is especially appropriate where winters are characterized by heavy snowfalls. Melting snow could easily result in waterlogged piles. Moreover, the snow would hinder the functioning of the handling equipment. In regions characterized by high prevailing winds, a windbreak on the windward side would be advantageous.

### *2. Setting Up the Windrows*

In conducting a windrow compost operation, material to be composted is piled in long windrows directly after the completion of the size reduction and sorting steps. The method of setting up a windrow should be adjusted to local conditions and to the nature of the material to be composted. This particularly holds true when additives (e.g., sludge or manure) are to be utilized. If no additives are involved, the course of action is straightforward in that the windows are set up immediately as the material leaves the grinder or classifier, as the case may be. If additives such as sewage sludge or other wastes are to be included, then one approach is to build up the windrow by alternating layers of refuse with the additive. Additive and refuse later are mixed together (i.e., blended) at the first turning of the pile. If turning is not the method of aeration, then the two components should be mixed prior to setting up the windrows. Of course, other ways of adding additives are available. One is to discharge sludge and refuse simultaneously through a common port. Various mechanisms are available for constructing windrows automatically. One such is an arrangement of conveyor belts, with the discharge end adjusted to the height intended for the completed pile.

The dimensions of a windrow are determined ultimately by its aeration requirements and by the efficient utilization of land area. Since the maintenance of aerobic conditions in a windrow is a function of adequacy of interstitial volume, the permissible dimensions of a pile ultimately depend upon the structural strength of the particles that make up the pile. The structurally stronger the predominant component, the higher can be the windrow without undue loss of interstitital volume. With municipal refuse, the height is on the order of 1.5 to 1.8 m. Another factor to consider with respect to height is that determined by type of equipment used to aerate the composting mass. Width has little effect on aeration other than, of course, to determine the ratio of surface area exposed to inward diffusion of air. Because the amount of inward diffusion usually is not appreciable or sufficient to warrant unusual measures, the width is that dictated by convenience. If other factors do not intervene, a width of about 2.4 m is suitable. All dimensions could be expanded during winter to enhance the windrow's self-insulation capacity, that is to promote retention of heat, or in windy and arid regions to minimize moisture loss through evaporation.

The geometry of the pile should be based on climatic conditions and especially on the type of equipment used. If rains are frequent or heavy, the top should be conical in cross section so as to shed water. On the other hand, a flattened top is appropriate if rainfall is not a problem, because heat loss is reduced with such a configuration, and a greater proportion of the material is exposed to the high temperatures prevailing within the pile. The contour imparted by the aeration equipment probably is the ultimate deciding factor, inasmuch as altering the configuration would require an additional operation step, and thus would increase the cost of the process.

*3. Aeration*

A windrowed mass of composting material can be aerated by one or a combination of two methods, namely turning and forced aeration. The efficacy of turning has been proven by long experience, whereas that of forced aeration remains to be demonstrated for many types of applications.

***a. Turning***

In the turning operation, a pile of composting material is torn down and then reconstructed. The reconstructed pile either can occupy its original position, or one immediately adjacent or somewhat removed from the original position. The act of tearing down and reconstructuing results in the exposure of the composting material to the ambient air as well as the renewal of the interstitial air, and thereby in the replenishment of the oxygen supply. Since the purpose of turning is to renew the oxygen supply, it should be done in such a manner as to avoid compacting the composting mass.

Ideally, when reconstructing the pile, material that had constituted the inner portion of the original pile should become the outer layer of the reconstituted pile. The rationale for the reversal of positions is twofold: (1) promotion of uniform decomposition; and (2) exposure of all material to the high temperatures that prevail in the interior of an actively composting pile. The exposure is an important factor in the destruction of disease-causing organisms that are present in the wastes.

Frequency of turning obviously has a determining effect on design and size of equipment utilized in the accomplishment of turning. Hence a few words on frequency of turning are appropriate at this time. Since aeration is the purpose of turning, the appropriate frequency depends upon the rate of oxygen uptake by the microbial population in the pile. Experience has shown that oxygen requirements in composting municipal refuse can easily be met, that is, maintained at optimum concentration, by turning the pile every third day, providing the material is not waterlogged or compacted.[1] The latter two can be remedied by increasing the frequency of turning.

A complicating factor other than aeration may result in a need for a more frequent turning. It pertains to the hygienic aspects of composting, especially in those operations in which sewage sludge is combined with refuse. Results of University of California studies[2] indicate that turning three times per week was not sufficient to bring about a complete die-off of enteric indicator organisms. This is not surprising in view of the fact that at peak activity, only about 40 to 50% of a typical windrow is exposed at a given interval to temperatures lethal to disease-causing organisms. Recontamination of the inner material "sterilized" by exposure to the high interior temperatures by the nonsterilized outer portion occurs during the turning operation. The recontamination may be counteracted by a more frequent turning program.

To allow for an increased frequency dictated by emergency situations (e.g., excessive moisture) and to ensure hygienic safety, the equipment capacity should be such that the composting material could be turned once each day if the need should arise.

FIGURE 1. Turning a pile with a front-end loader.

***b. Equipment***

Other than the size reduction and the sorting steps, turning can be the most highly mechanized operation in municipal windrow composting. In early practice, turning was done simply by spreading and restacking the material with a bulldozer. A later development consisted of doing the same task with the use of a front-end loader. This practice still persists, especially in small-scale operations. If a front-end loader is used the operator should be instructed to empty the bucket in a manner such that the contents cascade down rather than drop as a single mass. The operation is shown in Figure 1. Both methods are makeshift and unsatisfactory because the composting material is compressed rather than fluffed. Moreover, the operation becomes inefficient due to limitations with respect to the capacity of the machines to handle the large volumes of material to be processed, even in modest scale operations (e.g., 91 Mg/day).

Although as long ago as the early 1950s the mushroom industry had an effective automatic turning machine designed specifically for composting, it was only within the past 10 years that efforts were made to design machinery for municipal size operations. (Actually, because of the dearth of operations in the U.S., machines that have been built have seen service almost exclusively in composting feedlot manures and agricultural wastes, and to some extent, sewage sludge.) Of those on the market, five come to mind: the Scarab,* the Cobey,** the Terex,*** and the Brown Bear.**** The mechanisms by which the machines accomplish the turning differ among themselves. For example, with the Terex,® turning is accomplished by means of a deflector and plow. With the Cobey,® plow fenders gather in and guide the wastes towards a hydraulically driven drum equipped with rows of hardened teeth and rotated at a very high speed. The drum picks up the material and drops it in back of the machine. The configuration of the teeth is such that the processed material is deposited to form a pyramid-shaped row. The Brown Bear® does its stirring by means of a 3 to 4-m wide augur, 1 m in diameter. The Scarab® forms windrows 1.5 m high and about 4.2 m wide. It straddles the pile, and stirs and agitates the material in place, that is, the windrow remains in its original position. Its agitation mechanism is shown in the diagrammatic sketch in Figure 2. On the other hand, the Terex® and the other three mechanisms displace the original row to a position alongside the machine.

When consulting the record of a machine used in processing sludge and sawdust or agricultural wastes to evaluate its capacity to handle municipal refuse, it should be remembered that municipal wastes differ from the other materials drastically in terms of particle size distribution, texture, and handling characteristics. The differences undoubtedly are reflected in the performance of a particular turning machine. With this allowance in mind, one can expect a probable turning capacity of about 900 Mg/8-hr day.

### c. *Forced Aeration*

The possibility of aerating a mass of composting material by forcing or pulling air through it has intrigued designers of compost systems since the time municipal scale composting was first proposed in this country. If it could be made to function successfully, forced aeration would have many attractive features, not the least of which would be simplicity of equipment and operation and consequently a sharp reduction in costs.

Arrangement of a windrow to allow for forced aeration can be made in various ways. One method is to construct the pile on a somewhat elevated perforated base through which (i.e., through the perforations) air can be forced into the pile. Another arrangement is to install perforated ducts in a bottom layer of material having the consistence of sawdust or wood chips, and then piling the material to be composted on top of the layer. This latter is a key feature of the Beltsville aerated-pile method developed by G. B. Wilson and Epstein and colleagues.[3,4] Innovations introduced in the Beltsville system are (1) initially pulling air through the pile and later pushing it through; (2) discharging air exhausted from the composting material through a smaller mass either of fully composted material or of sawdust; and (3) encasing the pile with a 30-cm layer of composted material, wood chips, or comparable material to insulate the pile and thereby ensure the simultaneous exposure of all "fresh" material to high temperatures. Step 2 was introduced to remove objectional odors that might be in the discharge gases. An indication of the required air flow may be found in the fact that in one of its studies,[3] the U.S. EPA observed that an air flow of about 5 $m^3$/min at the blower inlet was sufficient to provide an oxygen concentration of 5 to 15% in a windrow of about 40 Mg of combined sludge cake and wood chips, the dimensions of which were $12 \times 6 \times 2.5$ m high. It should be emphasized that a 5-$m^3$/min air flow is by no means universally applicable. The required rate varies with each operational set-up.

* Scarab Manufacturing and Leasing Inc., White Deer, Texas, 79097.

** Cobey Composter, Division Eagle Crusher Co., Inc., Route 2, Box 72, Gordon, Ohio, 44833.

*** Terex, Terex Division of General Motors, Hudson, Ohio, 44236 (as of 1973).

****Brown Bear; Roscoe Brown Corp., Lenox, Iowa, 50581.

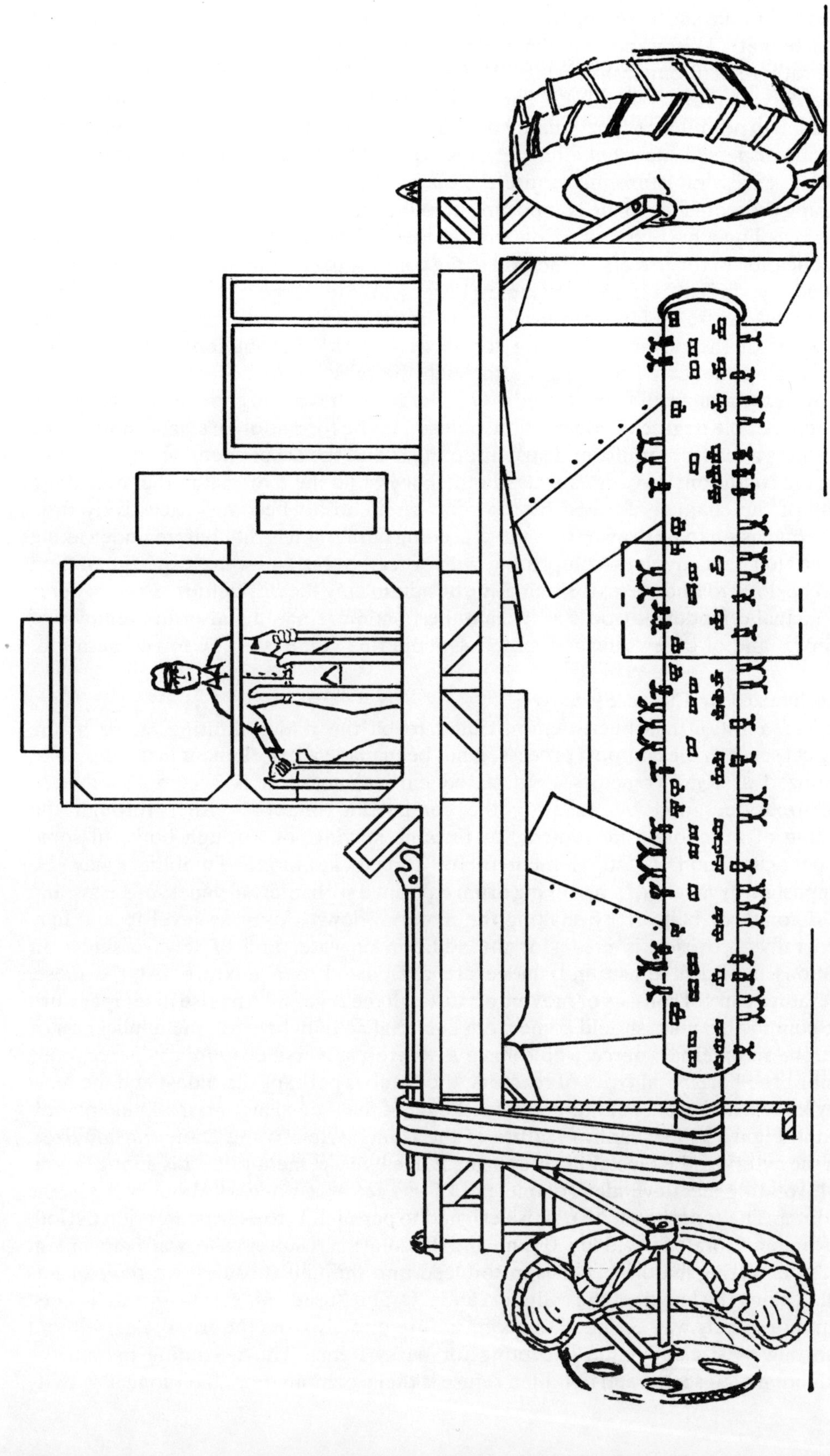

FIGURE 2. The Scarab® windrow turning machine.

Essential to the successful operation of a forced-air compost system is the utilization of a waste that: (1) is granular and noncompacted; (2) has a relatively uniform particle size of rather small dimensions (e.g., 1.25 cm); and (3) is not excessively moist (i.e., about 55% with municipal refuse). The granular nature and relatively small particle size are needed to permit air to pass unimpeded through the composting material, and yet be uniformly dispersed such that all material is exposed to the air flow. Aside from filling the voids, excess moisture may cause the material to compact. By relating the three preceding prerequisites to the composting of municipal refuse by the forced-air method and by recalling a past history of less than successful results, it becomes apparent that the application of the forced-air method to the composting of municipal refuse would be attended by difficulties. For example, because of its large percentage of paper, comminuted refuse is far from granular in consistency. Furthermore, because the dimensions of the particle sizes are too large and the distribution not sufficiently uniform, air is deflected as it passes through the mass, and short-circuiting ensues. Moreover, the structural strength and other characteristics of the paper is such as to lead to an appreciable degree of compaction, at least, to the formation of small clumps. The net result is a far from uniform distribution of air and the subsequent development of anaerobic conditions in various zones in the pile, while the material in the immediate vicinity of the channels formed by the short-circuited air becomes excessively dry.

The conclusion to be drawn from the preceding paragraph is that before undertaking an operation that involves composting with forced aeration, some experimentation should be done to find ways of eliminating or minimizing the difficulties. Since paper is the principal offender, it would seem that the fraction remaining after the removal of the paper, and of other salvaged materials, could be composted by forced aeration.

### C. Mechanized or Closed Systems

Because aeration has been demonstrated to be the major limiting factor in the biological sense in the compost process, it has been made a key element in the design of mechanized systems (reactors). As stated earlier, aeration is accomplished in a mechanized operation by agitating the composting material either through the imparting of a tumbling movement, or through stirring, or through both. In some reactions, agitation is supplemented by forced aeration. Tumbling may be accomplished by means of a revolving drum equipped with internal vanes, of a traveling endless conveyor belt, or by shaking the material down from one level to another.

Generally, provision is made for the addition of water and of sewage sludge in operations in which refuse and sludge are composted as a mixture. At the most, temperature control consists of moving air through the reactor. Any rise in temperature of the composting mass should come from bacterial activity because the application of heat from an external source would have a severely adverse economic repercussion.

Among the horizontal types of reactors, the Dano is perhaps the oldest and the most widely known and used. A diagrammatic sketch of the typical layout of a Dano plant is shown in Figure 3. The central feature of the Dano system is the Dano biostabilizer. From the exterior, the biostabilizer appears much like a cement kiln. It is a long (about 25 m), rotating steel cylinder (diameter, 3.5 m) set at a slope of about 5° from the horizontal. The capacity of the unit is designed to permit a 1- to 3-day detention period. The cylinder is rotated at 1.5 to 5.0 rpm, and the rotation results in a forward movement of the composting material. Air is introduced into the unit through two rows of jets installed along the length of the cylinder, and $CO_2$ and "used" air escape through vents. The unit originally was designed to accept refuse directly from the truck, i.e., without any sorting or size reduction excepting for bulky items. The reasoning behind the introduction of unsorted and unmilled refuse is that a combination of biological activity

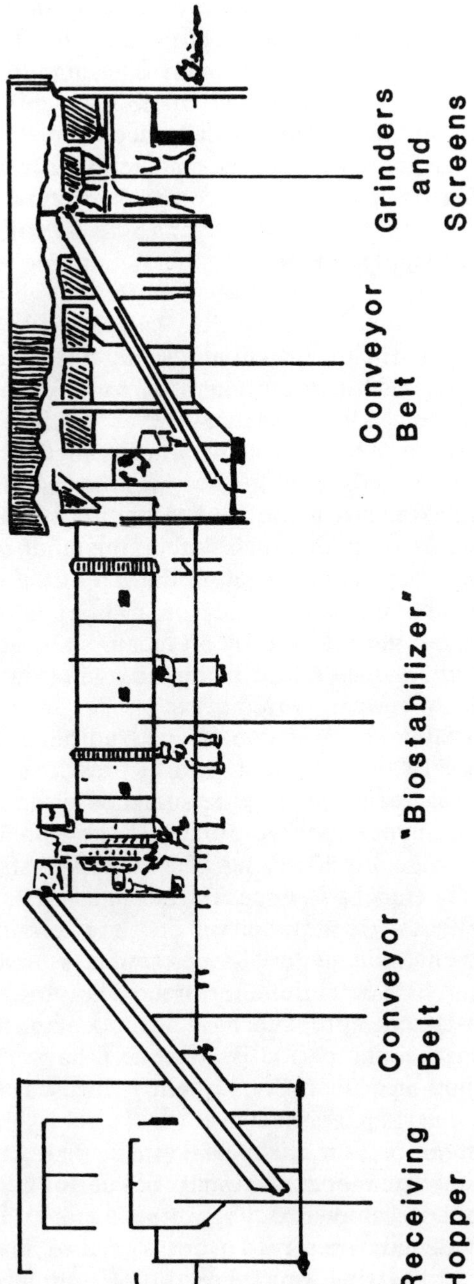


FIGURE 3. Layout of a Dano compost plant.

and the abrasion of the rotating material brings about the disintegration of the particles. Cans and other noncompostable items were removed after the composting material was discharged from the biostabilizer. It should be emphasized, however, that the successful operation of a Dano composter with refuse of the type produced in the U.S. demands the milling of the refuse prior to its introduction into the unit.

The brief sojourn of the composting material, even though sorted and milled, in the biostabilizer is far from sufficiently long to permit the extensive biological decomposition needed to reach stability. What really is accomplished by the action of the unit is a thorough preliminary aeration and possibly a stabilization of the small portion of the refuse that may have been partially decomposed at the time it was introduced into the drum. The greater part of the mass requires further decomposition and stabilization after leaving the drum. This is accomplished by stacking the material in windrows and allowing it to "cure" or "mature" over a 5- to 6-week period. Because of the system ventilation in the cylinder and the continued tumbling action, nitrogen loss by way of $NH_3$-H is greater with a biostabilizer than with a windrow.

Another common configuration is the vertical silo digester. The silo may be clindrical, square, or rectangular in cross section. Usually it is divided into a series of levels or floors. With one system, the floors consist of rows of troughs, V-shaped or trapezoidal in cross section, that can be inverted. Another arrangement is one in which the material on each floor is stirred by a revolving arm attached to a central shaft that extends the full height of the silo. During the stirring, the material is moved to an opening in the floor through which it is dropped to the floor below. The detention period on each floor is about 24 hr. In one such system, the number of floors is six. In vertical systems, the prepared refuse is inserted at the top of the reactor and then is slowly moved downward. Detention times range from as brief as 1 day to as long as 6 days. As with the Dano system the refuse is by no means stabilized by the time it is discharged from the digester, and therefore it must be windrowed and allowed to "cure" or mature over a 5- to 6-week period.

A system that first appeared under the name of The Naturalizer® System, later as IDC® (International Disposal Corp.) System, and at present as the Real Earth® System, centers around the use of a plug-flow reactor housed in a vertical building equipped with horizontal, continuous conveyors on which the ground refuse composts. The conveyors are about 3 m wide and 50 m long. The material is stacked to a depth of about 1.8 m on the belts. At the end of a 24-hr period the material on one belt is dropped or transferred to a lower belt. A system of fans provides the aeration. Usually after 2 days of exposure on the belts, the material is ground a second time, and then reintroduced into the digester. The detention time in the digester formerly was set at 5 to 6 days. However, recently the entrepreneurs have been speaking in terms of a 1- or a 2-day detention period. Whether the period is 1 day or 6 days, the material is not sufficiently stabilized and is poor in quality. Consequently, the usual 5- to 6-week period of maturation is required.

The Metro-Waste® System is a mechanical system that has been operated successfully in terms of technology. Horizontal tanks with perforated bottoms serve as reactors in the system. Air can be blown through the bottom either periodically or continuously. A specially designed agitator-unloader is used to mix the material and thus augment the aeration supplied by forced aeration. Essentially the agitator is a traveling endless conveyor belt mounted on wheels that move on tracks placed at the bottom of the tanks. The ground refuse is kept in the tank from 4 to 6 days. Thereafter it is stacked in windrows to cure over a 5- to 6-week period.

The last type of mechanical composter to be described herein is the Fairfield-Hardy® digester. The digester is a circular shaped vessel built of masonry or steel with a structural steel carriage pivoted in the center. Drive wheels located at the perimeter end of the carriage ride on a track around the top of the circular wall. Multiple augers supported from the carriage extend down into the reactor tank. Ground refuse is introduced along the inside wall of the reactor by integral conveyors. The material is aerated and moved towards the center discharge by the action of the multiple augers. Air is forced into the reactor by a motor-driven blower, and automatically metered into air pipes having discharge holes to distribute the air as required throughout the lower layer of material inside the tank. A distributing conveyor moves incoming material from the center inlet hopper to a chute immediately inside the wall at the top of the tank. The conveyor is mounted on top of the pivoted carriage and distributes the incoming material evenly along the inside circumference of the tank as the carriage and its multiple augers move around the track. The center discharge opening can be telescoped and thereby provide a means for regulating the depth of the digester contents. The relation of the various components to each other is shown in Figure 4.

## III. OPERATIONAL PARAMETERS

Certain important operational parameters are common to both windrow and mechanical systems. The parameters are (1) temperature rise and fall; (2) change in odor and color; (3) oxygen uptake; (4) moisture content; (5) pH level; and (6) degree of stability.

### A. Temperature

One of the more obvious manifestations of microbial activity in the compost process is the rise in temperature that begins almost immediately after the prepared material is windrowed or placed in a digester. The rise is the result of energy generated in bacterial metabolism over and above that used in cellular synthesis. Microbes, in common with all living beings, do not attain 100% efficiency in the utilization of energy produced in the metabolism of carbon. The energy not used in the cell is released in the form of heat. If the heat cannot escape or escapes only in part, the temperature of the composting mass rises. In a windrow the mass of material serves as insulation to retain the heat. The phenomenon also applies to enclosed composting. It follows, therefore, that the total amount of heat released, and therefore the extent of temperature elevation, is determined by rate of bacterial activity. In the early stages of the process, when the readily decomposable material is being broken down at a relatively rapid pace, and therefore bacterial activity is at its peak, the temperature rise and fall is rapid and pronounced. As the readily decomposed material disappears, microbial activity lessens, and consequently, the temperature begins to drop. Eventually, all material susceptible to microbial attack will have been decomposed at a rate slow enough not to result in an accumulation of objectionable intermediate and end-products. At this point, the temperature will have dropped until that of the ambient temperature is reached. The more closely the end point is approached, the lesser will be microbial activity, and the lower will be the temperature of the composting mass.

From the preceding description of the overall reaction of the composting mass in terms of temperature, it can be seen that if the course of temperature rise and fall were plotted, the shape of the resulting curve would be closely analogous to that of a typical growth curve. The resemblance is due to the fact that the compost temperature curve represents an integration of the many growth curves of the succession of microbial populations that participate in the process. Experience has shown that after the final temperature decline has reached 40 or 45°C, the material is sufficiently stable to be stored without danger of becoming a source of nuisances.

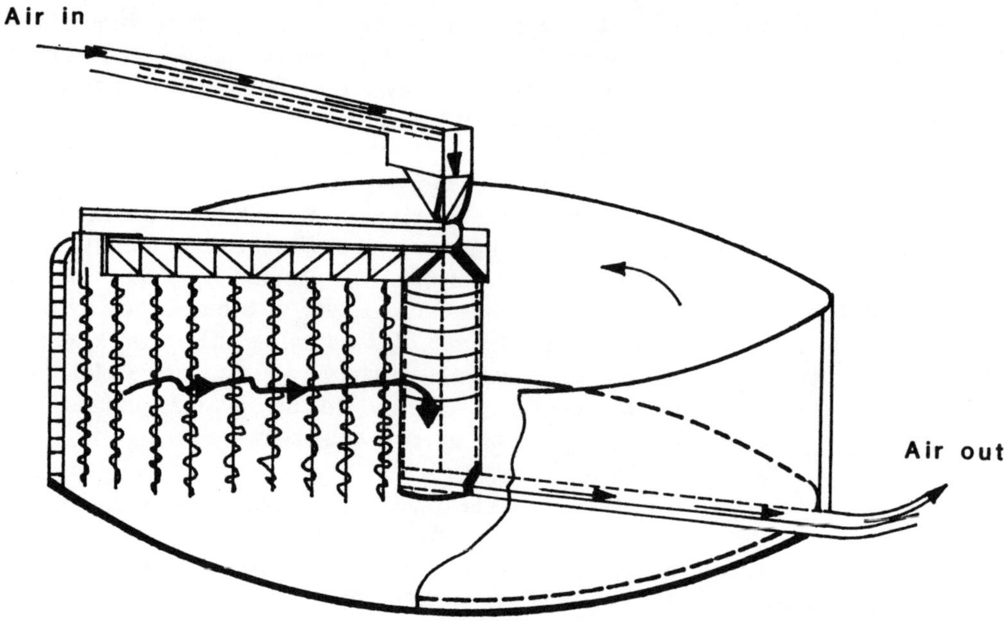


FIGURE 4. Fairfield-Hardy digester.

An interruption or inhibition of the bacterial activity is manifested by a departure from the normal temperature pattern. It follows that the extent and duration of the departure depends upon those conditions that brought about the inhibition. Consequently, temperature becomes an excellent operational parameter. (However, it must be emphasized that the attainment and maintenance of a high temperature does not preclude the existence of foul odors.) The two factors most commonly responsible for abnormal drops in temperature are an unfavorable moisture content and an inadequate supply of oxygen. The normal final decline in temperature can be recognized by the persistence of the drop in the face of optimum operational conditions.

**B. Color and Odor**

Change in color and odor may be regarded as qualitative indicators of bacterial activity in that it represents a change in the nature of the wastes. Color of the material shifts from the grayish-green of freshly milled refuse to a dark gray with brown overtones within a few days after the process has been initiated. The odor is transformed from an originally pungent to a somewhat musty, earthy odor. The latter stems from the presence of actinomycetes and fungi. Odor can be used as a crude indicator of the development of anaerobic conditions, since a characteristic of the latter is the production of very foul stenches. Therefore, with a windrow operation, the onset of objectionable odors should be a signal for increasing the frequency of turning and for checking for excess moisture. If the anerobiosis stems from excessive moisture, turning the pile more frequently to accelerate evaporation of water will remedy the situation.

If time and land area are not of the essence, an alternative to increased frequency of turning is not to turn the pile over a 30- to 60-day period. The rationale for so doing is twofold. First, the release of odors from the pile is far less extensive than that which occurs during the turning operation; and secondly, decomposition, even though largely anaerobic, continues to take place, and chances are that the material will have been sufficiently stabilized to result in a substantial lessening of odors by the time of the next turning. A "bonus" is a lessening of the loss of nitrogen. However, aside from the disadvantages attending the prolongation of the detention period, others of a rather serious nature may arise. For example, the undisturbed piles not only may attract flies, they also may serve as a source of nutrient and provide the means for fly development and propagation. Moreoever, the problem of odor emission is not eliminated—the immediate vicinity of the pile will be permeated with an objectionable odor.

With an enclosed digester, the development of foul odors indicates the need to increase the air flow through the unit and to agitate the reactor contents more vigorously.

### C. Oxygen Uptake

As would be expected, the rate of oxygen uptake is a useful indicator, inasmuch as aerobic microbes constitute the major biological agents in the compost process. Because of the latter fact, rate of oxygen uptake is a measure of biological oxidation and hence of decomposition of the waste. However, because of the nature of a windrow, the oxygen concentration in it cannot be measured precisely. On the other hand, the odor parameter described in the preceding paragraph is satisfactory for routine operations. A disadvantage in relying upon foul odors as a parameter is that they fall into the category of "after the fact" phenomena. Naturally a more satisfactory parameter would be one that anticipates the development of an undesirable situation.

The determination of oxygen concentration in an enclosed unit is relatively straightforward and can be made by measuring the oxygen content of the discharge air stream. As a precaution, the air input should be such that the discharged air contains some oxygen. The actual percentage oxygen content of the discharge stream consonant with the maintenance of aerobiosis in the composting mass varies with type of equipment and the characteristics of the composting mass, and hence must be established for each operation. The same reasoning applies to the rate of air input required to ensure a sufficient supply of oxygen in the unit. While a considerable amount of research has been done on oxygen uptake and required air flows,[5-7] the fact remains that the number of numerical estimates matches that of the researchers. The variation in numbers confirms the need for a specific determination of the air flow rate required for a given operation.

The following examples provide an approximation of the $O_2$ concentrations involved:

1. An $O_2$ pressure greater than 14% of the total indicates that not more than one third of the $O_2$ in the air has been consumed.
2. The optimum $O_2$ level is 14 to 17%.
3. Aerobic composting supposedly ceases if the $O_2$ concentration drops to 10% by volume.
4. If $CO_2$ concentration in the exhaust gas is used as a parameter for $O_2$ concentration, then the $CO_2$ in the exhaust gas should be between 3 and 6% by volume.

### D. Moisture Content

The importance of moisture content and the appropriate levels thereof were discussed in the section on environmental factors. Here it is considered in its function as an operational parameter. The most obvious indication of excessive moisture is the production of leachate (seepage) from a windrow, and in a mechanized digester, aggravated clumping or balling. A symptom equally obvious in its manifestation is the development of the foul odors described earlier. It goes without saying that such indicators are too crude for application in an efficient operation.

The most accurate method for determining moisture content continues to be the one in which weight loss resulting from exposure to a given temperature (e.g., 103°C) in a conventional laboratory oven over a defined time period (2 to 6 hr) is determined. The main drawback of this method is that it is time consuming. A method equally precise and accurate has been developed by Cardenas, Jr.[8] which involves the use of a microwave oven in which exposure time is reduced to 15 to 20 min.

### E. pH

As stated earlier, the pH level of properly composting material rises slowly. A departure from the slow rise represents a disturbance of the compost process. If a deviation should occur, it probably would be in the form of a drop in pH level due to acid formation brought about by anaerobic conditions. If the latter were true, then the other symptom (i.e., foul odors) of anaerobiosis should also be noticeable. Accordingly, the remedy would be stepped-up aeration.

### F. Stability

The last parameter to be discussed, namely degree of stability, is one for which a satisfactory analytical method remains to be developed. Indeed, if a reliable analytical technique for determining degree of stability could be developed that were truly universal in its application, it would be a great boon to practical composting. The reasons are many. It is an important parameter because it determines when the "active" phase of the compost process has been completed, and hence the time after which no further handling is necessary other than preparation for storage or use of the product. The parameter is economically significant because it places an upper limit on the required equipment capacity as well as determines the extent of labor, power, and maintenance expenditures. An added utility would come from its making possible the establishment of a standard for evaluating the efficiency of the various compost systems on the market in terms of time required to attain stability. Such a standard would be especially valuable in deflating exaggerated claims advanced by an over-zealous entrepreneur. As an "aside", it should be pointed out that unless one is dealing with an unusually "unstable" (easily decomposed) waste, and municipal waste does not merit such a qualification, claims of compost times of less than 3 weeks for municipal refuse, *inclusive* of curing or maturation should be viewed with a high degree of skepticism.

Although color and odor changes are useful indicators of microbial activity, they are not adequate for basing a decision as to the maturity of a composting material. In reality, change in both occur long before stability is reached. This latter fact can lead to unsatisfactory practice and exaggerated claims. Unsatisfactory practice consists of prematurely sacking or spreading the composting material on the land. Exaggerated claims come through the unrealistic attribution of rapid composting to a system solely on the basis of color and odor as indicators of stability.

The following is a list of methods proposed for determining degree of stability: final drop in temperature,[1] degree of self-heating capacity,[9] amount of decomposable and resistant organic matter in the material,[10] rise in the redox potential,[11] oxygen uptake,[12] growth response of the fungus *(Chaetomium gracillis),*[13] and the starch test.[14] Of the tests, the final drop in temperature probably is the most reliable, inasmuch as it is a measure of total microbial activity as well as of rate. The measurement pertains to rate because it is an expression of the difference between total heat generation and rate of heat loss. If bacterial activity has lessened, the rate of heat generation is correspondingly lowered, and hence the differential between heat generation and heat loss (heat loss remaining relatively constant) becomes less, and a lower temperature of the mass is the result. It has the qualification of universality inasmuch as it is a measure of rate of bacterial activity directed towards the breakdown of organic matter. The disadvantage with decline in temperature as a parameter is the time element. Inasmuch as the decline represents a trend, it implies a number of readings taken over a period of time.

The remaining six tests do not meet the requisite of universality of application. By this it is meant that the numerical values obtained in the tests have significance only with respect to the specific material and compost procedure for which the tests were made. For example, a redox potential that characterizes stability under one set of compost conditions does not necessarily do so under another set. The problem of lack of universality with some is aggravated by the difficulty of conducting the test itself. The *Chaetomim* test is an example.

### G. Recent Advances in Engineering Research

After reading the preceding sections on operational parameters, the reader could justifiably conclude that the measurements and specification of design and operational parameters for composting are largely empirical in nature. However, this situation now is changing due to a renewal of research effort that is accompanying the resurgence of interest in composting brought about by the widespread acceptance of the process in the treatment of wastewater sludges. Because the recent researchers mainly are in the ranks of engineers concerned with wastewater treatment and management of sludges, and because composting on an appreciable scale in the U.S. is confined to that of sewage sludge, published reports of the new research deal almost entirely with the composting of sewage sludge. The method ususally described is the aerated static-pile method.

In their effort to advance composting from an empirical to a rational status, the researchers began by attempting to develop and demonstrate a theoretical basis for the process. For example, through the development of heat and mass balances, Haug[15,16] demonstrated the existence of a critical ratio of water-to-biodegradeable volatile solids, and that in the process, composting and drying are sequential operations. Finstein,[17] concerned with the relationship between microbiological activity and temperature, related maximum heat production and accompanying water loss to the control of the temperature by means of rate of aeration. Halley, Shea, and Braswell[18] demonstrated the significance of the specific aeration factor ($m^3$/min/kg dry solids). The upshot of these recent research efforts is a reemphasis of the importance of aeration rate as a means of optimizing selectively the biological decomposing and drying of the composting material within the period of active aeration.

## IV. ECONOMICS

### A. Introduction

Total cost figures given for compost operations in the literature almost always include those for sorting and size reduction. This inclusion is justified only if composting is the sole form of resource recovery being carried on, or if it is the only one requiring sorting and size reduction. Since in terms of capital and operation the costs of these two preliminary steps constitute a significant fraction of the figures quoted in the literature, the costs of composting would be substantially lowered in an operation in which composting is but one unit process in a total resource recovery system. Naturally, in such a case the costs of sorting and size reduction would be prorated among all unit processes that require such treatment.

Another observation to be made is that factual monetary values are exceedingly difficult to find in the literature. Generally the available cost figures are those obtained from promotional literature or from interviews with biased parties. As such, one would be safe in regarding them (i.e., the quoted numbers) as being unrealistically low.

However, a few attempts have been made to arrive at realistic numbers. The results of one such study are contained in two reports issued by the U.S. Environmental Protection Agency on its demonstration project at Johnson City, Tennessee.[19,20] Dollar values given herein for windrow composting are taken in large part from the two publications. An approximation of 1979 dollars can be made by multiplying the given dollar values by 1.92.[21]

## B. Capital Costs

As is to be expected, the most realistic numbers given the EPA reports are those pertaining to the demonstration plant used in the agency's Johnson City project. Inasmuch as full details of the construction and operation of the plant are given in the two reports mentioned in the preceding section[19,20] for the purposes of this publication, it suffices to state that a windrow system was used. The capital costs of the Johnson City plant and estimated capital costs for windrow plants of varying capacity are presented in Table 1, which is reproduced from the second EPA-TVA report.[20] Because the projections are based upon costs at Johnson City, actual costs would vary at other locations. This would be especially true with site improvement and land costs. Other variations would arise from the incorporation of different types of size reduction and other equipment. With the Johnson City plant, size reduction and sorting equipment accounted for about 13% of the capital cost. The data in the table indicate the existence of a pronounced effect of economy of scale on capital cost per Mg of daily capacity. On the basis of the figures, the total cost per Mg of refuse processed in 1969 ranged from \$11.81 (180 Mg/day) to \$21.79 (45 Mg/day).

The economy of scale undoubtedly continues beyond the 180-Mg/day plant (90 Mg/8-hr shift). For example, the average turning machine can turn from 720 to 900 Mg of material per hour, and therefore one machine is sufficient for a 720- to 900 Mg/day plant. Thus, with respect to turning equipment, the economy of scale extends to a 720- to 900-Mg/day capacity. The capital cost of an automated turning machine (Terex 74-51)® is about \$311/Mg (1969 dollars) based on a single 8-hr shift.[22] Total ownership and operating costs would be less than \$1/Mg. The upper level of capacity at which the breakpoint in economy of scale occurs remains to be determined.

A source of information on capital costs for mechanical compost plants is a paper by C. I. Harding.[23] The data in Table 2, taken from Harding's article, give an approximation of the capital costs of three mechanical systems tried in the U.S. Although the dollar values listed in the table apply to 1968, they can be readily extrapolated to approximate 1979 values by multiplying by 1.92.[21] Thus, extrapolating Harding's numbers to 1979 values, the capital cost per megagram daily capacity for either the Fairfield-Hardy® System or the IDC® System would range from about \$17,067 (360 Mg/day) to \$29,867/Mg (90 Mg/day). An estimate based upon one source and extrapolated to 1979 dollars places the capital cost of the Metro® System at \$14,720/Mg daily capacity with a 324-Mg/day plant.[24] An estimate for the IDC (Naturizer) system as based upon another source[25] and extrapolated to 1979 dollars comes up with a capital cost of \$24,490/Mg for a 135-Mg/day plant.

## C. Operating Costs

The two U.S. EPA reports cited for capital costs[19,20] are also excellent sources of data for operating costs with windrow systems. Cost estimates which include those for the complete operation, those exclusive of sewage processing, and finally those exclusive of sorting, grinding, and sewage processing are given in Table 1. The projections are based on the Johnson City demonstration project and probably are conservative. Another point to keep in mind is that the economy of scale undoubtedly extends beyond the 90-Mg/day (two-shift) capacity.

**Table 1**
**CAPITAL OPERATING AND TOTAL COSTS FOR WINDROW COMPOSTING[a]**

| Plant capacity (Mg) Day | Year | Shifts (no.) | Capital cost[b] ($/Mg/day) | Capital and investment $/Mg | Operating[c] $/Mg | Total $/Mg |
|---|---|---|---|---|---|---|
| 47 | 12,260 | 1 | 20,484 | 7.59 | 14.78 | 22.36 |
| 45 | 11,790 | 1 | 16,974 | 6.75 | 15.05 | 21.79 |
| 90 | 23,580 | 2 | 9,404 | 3.47 | 11.58 | 15.05 |
| 90 | 23,580 | 1 | 11,047 | 4.06 | 10.91 | 15.07 |
| 180 | 47,164 | 2 | 6,019 | 2.22 | 9.59 | 11.81 |

[a] The data in the table are based on information from U.S. EPA report.[20] Amounts based on actual costs of Johnson City composting plant with modifications with 7.5% financing over 20 years; straight line depreciation of buildings and equipment over 20 years.
[b] Includes land costs estimated at $1960/ha.
[c] Does not include costs for landfilling rejects.

**Table 2**
**ESTIMATED CAPITAL COSTS OF THREE MECHANICAL COMPOST PLANTS[a]**

| | Fairfield | | | Metro-waste | | | IDC[b] | | |
|---|---|---|---|---|---|---|---|---|---|
| Capacity (Mg/day) | Cost ($ × 10⁶) | Power (kW) | Laborers (no.) | Cost ($ × 10⁶) | Power (kW) | Laborer (no.) | Cost ($ × 10⁶) | Power (kW) | Laborers (no.) |
| 90 | 1.4 | 671 | 8 | 0.9 | 932 | 12 | 1.4 | 447 | 20 |
| 180 | 2.1 | 1043 | 11 | 1.2 | 1268 | 17[c] | 2.1[c] | 596[c] | 28[c] |
| 270 | 2.5 | 1268 | 14 | 1.5 | 1417 | 25[c] | 2.7[c] | 708[c] | 36[c] |
| 360 | 3.2 | 1864 | 20 | 1.6 | 1491 | 30[c] | 3.2[c] | 820[c] | 45[c] |

[a] Exclusive of cost of land and special foundation problems (fill or piling, or both).
[b] IDC® originally was the Naturizer® process, and presently is the Real-Earth® system.
[c] Harding's[23] estimate as based upon chemical engineering estimation procedures.

Judging from available promotional literature, the operational costs for mechanized systems probably are comparable to those for windrow composting. It should be remembered that when considering the figures on operational costs (as well as those on capital costs) proffered by entrepreneurs for closed compost systems, past experience reveals a tendency on their part to underestimate operational requirements and to underdesign equipment. Underdesigning of equipment leads to a high rate of breakdown and malfunction. The failing with respect to operational requirements usually is in terms of manpower. The outcome of both shortcomings is an unrealistic and overly optimistic cost estimate.

An estimate of a part of the operational cost can be made on the basis of the power and labor requirements proposed by Harding[23] (see Table 2) and a value for power reported by Gotaas for the Dano® System.[26] Gotaas names a power requirement of 12 to 16 kWh/Mg of refuse processed. According to Table 2, the IDC® System has the lowest power requirement of the three described by Harding. However, the power requirement for the Metro® System could be lowered somewhat if the bottom aeration of the compost troughs were eliminated. (It is very likely that the forced aeration in the Metro® process is unnecessary.) The choice between the windrow and the closed system of composting should be based principally upon economics, inasmuch as the two approaches are quite similar in terms of performance and environmental impact, and the total time is about the same with both. The brief detention periods in the digester unit in closed systems is counteracted by the long period of curing or maturation in windrows required after the composting material has been discharged from the reactor. "Curing" or "maturing" are but other names assigned to a process that essentially is composting. ("Curing" corresponds to the period of plateau and final drop in temperature observed in windrow composting.)

As far as environmental impact is concerned, the simple act of enclosing (i.e., mechanizing) the compost process is no insurance against an unfavorable impact. In fact, each of the few large-scale operations that had been in existence in the U.S. at one time or another was threatened with enforced closure by local health authorities because of nuisances, primarily in the form of odors emanating from the compost plant.

## D. Market

### *1. Value of the Compost Product*

The potential market for a product is a function of the utility of the product to the prospective user. Composting and its products are no exception. With compost, one must distinguish between its intrinsic worth and its actual monetary value as determined by extent of the demand for it. A problem arises when dealing with the second part of the distinction.

#### ***a. Intrinsic Worth***

The utility of the compost product is in its function as a soil conditioner and secondarily as a mulch. Of course, the latter eventually becomes incorporated into the soil and then acts as a soil conditoner. When incorporated into the soil, compost continues to decay. Through the decay, organic intermediates are formed that change the texture of the soil such that clay soils are "loosened" and the sand particles in sandy soils are aggregated. The result in both cases is a "crumbly" soil. A consequence of the improved texture is an enhancement of aeration, since a crumbly texture facilitates the penetration of air. Perhaps of equal importance is the increase in the water-holding capacity of the soil to a degree that rivals that attained with the incorporation of peat moss. Furthermore, because it is an organic decomposition product, compost is effective in immobilizing heavy metals in such a manner that plants do not assimilate them, nor are the metals leached to the ground water.

In contrast to its favorable soil conditioner aspects, composted municipal refuse has a rather low fertilizer value since its combined nitrogen, phosphorous, and potassium (NPK) content usually is less than 3%; e.g., nitrogen from 0.5 to 1.57%, phosphorus from 0.16 to 0.4%, and potassium from 0.21 to 0.40%. A combined composting of refuse and sewage sludge ("cocomposting") may raise the NPK content somewhat, but not to a level that would permit it to be legally labeled a fertilizer. However, the presence of compost in the soil results in a more efficient utilization of inorganic fertilizer elements that may be added to the soil. The efficiency results from the binding of the elements in such a manner that they are not leached through the soil beyond the crop root zone. On the other hand, in the absence of compost, as much as 35% of applied inorganic nitrogen and 20% of applied phosphorus are lost through leaching.

Perhaps compost's greatest usefulness is in land reclamation. As such it can be used to at least partially restore areas ravaged by strip mining and fixing banks along highways and the spoils from other excavations. Its value in these applications is that it constitutes the humus needed to replace that lost when the soil was excavated.

Finally, it should be emphasized that the utility of the compost product is predicated upon its having a C/N no higher than 20. If the C/N is too high, the supply of nitrogen in the soil may not be large enough to meet the combined needs of the bacteria and the crop plants. Since the soil bacteria surpass the crop plants in efficiency of nitrogen assimilation, they use most of the nitrogen available, and as a result, the crop plants soon exhibit symptoms of nitrogen deprivation. On the other hand, if the C/N is too low, chances are that ammonia will be released into the soil. Ammonia is toxic to plant root hairs at all but the most minute concentrations.

### 2. *Demand for the Product*

In a study conducted on the potential demand for the compost product,[27] the compost product is classified into two main categories upon the basis of type of waste from which the compost was produced. The categories are (1) compost product from the organic fraction of municipal refuse combined with sewage sludge; and (2) a product which did not come from municipal waste, but instead was obtained either with somewhat segregated plant debris, or with manures, or with a combination of the two. The marketability of the two groups was explored in terms of potential large-scale demands down to those by home gardeners.

Generally, a certain reluctance was found at all levels to use a product obtained from municipal solid wastes and sludges. The reasons for the reluctance were not hard to ascertain. Thus, one objection was that the finished product would very likely contain bits and pieces of plastics, glass shards, and sundry other visually objectionable materials. The unpleasant visual impact of such contaminants drastically lessens the utility of the product for application in parks and recreational areas. It should be pointed out that the glass shards are not likely to lead to injuries in the form of cuts and abrasions, because sharp edges are polished (i.e., blunted) during the steps involved in composting. Although the contaminants might have no adverse effect on crop growth in large-scale agriculture, they could interfere with the performance of agricultural equipment. The home gardener is not happy with the contaminants because they are graphic reminders of the origin of the compost and they would litter his garden. A second reason for the reluctance is the concern about the possibility of disease-causing organisms and heavy metals being present in the compost.

The reluctance with respect to compost from municipal refuse and sludge is not evident with compost from plant debris and manures. These latter materials generally are free of the contaminants that detract from the eye appeal of municipal compost, as well as of disease-causing organisms and heavy metals. Moreover, they generally acquire the dark color and granular texture which are associated with a "good" compost. An added attraction is the fact that they have a higher NPK content than does run-of-the-mill municipal compost.

### 3. *Large-Scale Market—Problems*

There is no denying that the existing demand for compost in large-scale agriculture is minimal. The past and present lack of interest stems from a number of factors, of which the two major ones are (1) bulkiness of the product and all that bulkiness entails in the application of the material to the land; and (2) the present disparity between the dollar cost of chemical fertilizer elements and that of the compost fertilizer elements.

Because of its bulk, the utilization of compost involves the use of large-sized and generally heavy equipment. While it is true that heavy equipment may be used for harvesting crops, the use occurs at a time when the effect on the soil is least disastrous. On the other hand, compost would have to be applied at the time the soil is prepared and usually when the soil is wet. This latter results in the earth being compacted and thereby rendered unfavorable for crop production. While a solution would be to apply the compost directly after harvesting is completed, optimum crop production necessitates the application of the fertilizer elements during the growing season as well. The same reasoning applies to the use of compost as a mulch.

Perhaps a more significant factor is the disparity between the unit costs of fertilizer elements in the two sources (chemical fertilizer and compost). At present, the cost of chemical nitrogen, phosphorous, and potassium is only a fraction of that of the same elements in the form of compost. Moreover, of the two, the chemical fertilizer can be applied with far less effort and expense. Furthermore, since the weight and volume of the chemical fertilizer needed to supply an equivalent NPK is far less than that required with compost, it follows that transportation costs reflects this disparity. Moreover, because of proximity to the raw material, municipal compost plants would be near urban centers and hence at some distance from the sites of crop production.

One could surmise that if the disparity in each of application and unit fertilizer costs were compensated by an increase in yield, the reluctance on the part of large-scale agriculture would be overcome. Apparently, for the present this is not the case. In its Johnson City study,[20] researchers conducted several small-scale demonstration (field) studies in which significant increases in yield due to compost were observed. (A minimum of about 40 Mg compost/hectare was needed to bring about the increase.) Unfortunately, the dollar value of the increased crop production did not always match the expense involved in adding the compost. The exception was with tobacco production. Here, the high monetary value of tobacco made the increased production a monetary gain. Despite the higher yields, the farmers who took part in the study evinced little interest in continuing the application of compost on their own.

In interpreting the findings in the Johnson City study, it should be kept in mind that at the time of the investigation, chemical nitrogen was in abundant supply and hence was relatively inexpensive. Moreover, the monetary value of crops was much lower than it is today. With the dimunition of the availability of natural gas, the cost of chemical nitrogen is escalating. The upshot should be that the disparity in unit cost of nitrogen, of phosphorus, and of potassium between the chemical fertilizer and compost will lessen and conceivably even be reversed at some future date. However, that time apparently is not yet at hand.

An important factor that might help to break down the resistance to the utilization of compost in large-scale agriculture is an awareness of the increase in water-holding capacity of soil brought about by the compost. Despite the news of seasonal flooding in various parts of the nation, the well-high unanimous prognosis is for an increasing shortage of water in the face of the expanding needs of industry and agriculture. A consequence of the short supply is an inevitable rise in the cost of water. The dollar savings by way of water conservation made possible through the utilization of compost eventually may overshadow any disparity that may exist in unit fertilizer costs.

*4. Orchards and Vineyards (Luxury Crops)*

Utilization in orchards and vineyards is the largest potential outlet for the compost product. In the University of California study on the potential market for compost in California,[27] the researchers found that most grape and fruit producers return a sizeable percentage of prunings and crop residues to the soil. Moreover, pomace produced in the local wineries is made available to nearby producers. Thus, the growers apparently are aware that benefits are to be had by applying organic matter to the soil, and the transfer of this awareness to compost utilization should be only a small step. Judging from the study, about 10% of the growers in the Napa Valley* would use high-quality compost in their vineyards if it were available. Based upon an application rate of 44.5 Mg/ha and information from grape growers, approximately 27,000/Mg of good grade compost

*Napa Valley is located about 50 to 66 km northeast of San Francisco and is noted for its grape and wine production.

could be marketed each year in the Napa Valley (area, 607 ha). It should be emphasized that this potential demand is for a high-grade compost product. An important unknown is the amount of money the grower would be willing to pay for the product if he agreed to use it.

*5. Reclamation*

Reclamation may prove to be the principal use for municipal compost. Such a utilization has the advantage of not requiring a top-quality compost. In fact, about the major requirement is that the compost not pose a public health hazard. Consequently relatively low-grade municipal compost would be adequate. Since a low grade would suffice, less costly methods of producing the compost could be followed. That being the case, the method of choice would be the windrow method. As for the site of the operation, a study should be made as to whether the plant should be set up near the point of generation of the wastes, or in the vicinity of the land to be reclaimed. If composting at the site to be reclaimed proves to be the more feasible of the two possibilities, then the next determination would be on the point of removal of noncompostables, i.e., whether they should be removed before or after shipping.

Of course the monetary return per ton that could be expected from the sale of the product probably would be only slightly, if any, above the cost of production and transport. However, with even a slight return, when multiplied by the thousands of tons needed for strip mine reclamation, the dollar amounts could become substantial.

## V. POSITION OF COMPOSTING IN RESOURCE RECOVERY

### A. In Total Resource Recovery

Experience has fairly well shown that composting municipal refuse as the sole reclamation activity has yet to be proven economically viable in the U.S. In addition to the reasons adduced earlier is another factor peculiar to composting, namely, the application of a double standard when evaluating the process as a municipal waste treatment system. With other conventional and less conventional systems, success is judged only in terms of the performance of the systems. On the other hand, the evaluation of a compost operation includes not only performance but also economic success. Consequently if the compost product cannot be sold and therefore no monetary return is forthcoming, then the operation is considered to be a failure. This problem of the double standard would at least be minimized if composting were made but one part of a total resource recovery undertaking.

A more recent problem that militates against composting as the sole reclamation activity is a change in the composition of U.S. municipal refuse that has resulted in a rise in the C/N ratio of the organic fraction to 55/1 and higher. The implications of the high C/N, and of the choice between two courses to be followed to rectify it, were described earlier. Here, attention is called to the fact that in total resource recovery, a large portion of the paper would be salvaged either for paper manufacture or for use as refuse derived fuel (RDF), or for both. The consequence would be a lowering of the C/N of the residue intended for composting. Lowering the C/N would be but one of the benefits to be gained. Another would be the removal of the glass and plastic particles that would otherwise mar the compost product. The outcome would be a substantial upgrading of the product.

An extremely important reason for having a total resource recovery operation is that the costs of pretreatment (sorting and grinding) would be prorated among all the reclamation functions and would not be charged solely to composting. This would result in a substantial reduction in capital and operating costs. Moreover, the proceeds from the sale of the recovered resources could serve to defray the costs of the entire treatment process, or at least a part.

A trend responsible for the recent resurgence of interest in composting is the application of the composting process to sewage sludge disposal. At this time there is no need to dwell upon the many problems besetting the disposal of the increasingly large amount of sewage sludge being generated in the U.S. Suffice it to state that apparently for the present, a very attractive means of disposing of the sludge is to spread it on the land. The rationale is partly one of conservation, but mostly of economic expediency. A deterrent to this approach, however, is the possibility of its posing a health hazard. This is where composting comes in; namely, it is a means of minimizing or even removing that hazard. Other lesser reasons center around the improvement of the physical characteristics of the sludge.

As a consequence of the recognition of the potential of composting in making the sludge more acceptable for land disposal, a considerable number of studies ranging from small pilot to full demonstration scale have been and are being conducted. The outcome of the successes attained in the majority of the studies has led to the serious consideration of composting as an important process in sludge management.

Lest the reader leap to the conclusion that composting sludge is not attended by problems, it should be pointed out that certain troublesome questions remain to be answered. One of the more important is the extent to which disease-causing organisms are destroyed. As mentioned earlier, University of California studies on windrow composting of combined refuse and sludge[28] and those by others have shown that a high temperature alone, although important, does not account for the destruction of pathogens. Other factors such as antibiosis, competition, and destruction of nutrient collectively about equal it in importance. Undoubtedly, passage of time and careful monitoring are the most important safety factors.

An appealing approach to treating the two principal municipal solid wastes, i.e., refuse and sewage solids (sludge), would be to combine the less useful organic fraction of the refuse (e.g., unsalvageable paper) with the sludge and to compost the combination. The refuse could serve as a bulking agent and also act as a "diluent" for heavy metals in the sludge. The respective suitable ratios of the two would depend to a large extent upon the moisture content of the sludge. If sludge cake (approximately 15 to 20% solid) were used, a ratio of three parts refuse to one part sludge on a volumetric basis probably would be satisfactory.

Combined composting of sludge and refuse is not a novel concept. It has been practiced intermittently for many years in other countries, but mostly with the use of mechanized units, i.e., enclosed digesters. However, caution should be exercised in using this past experience as a guide. The reasons are (1) the operations were only grossly monitored; and (2) most had a history of becoming a source of nuisances. Whether or not the latter could be attributed to the sludge cannot be answered.

A major difficulty brought to light in the University of California study is that of mixing sludge cake with size-reduced refuse. In the process of mixing, the material has the unfortunate tendency to form balls or globs, the interiors of which soon become anaerobic, and hence serve as sources of foul odors. Work done by Cal Recovery Systems, Inc. shows that "balling" can be eliminated with the use of a properly designed mechanical turner.[29] The work also showed that turning a sludge-refuse mixture by means of a front-end loader can lead to the development of extremely foul odors.

In closing the section on composting, it may be conjectured that the course to follow in practicing municipal composting is not only to make it a part of a total resource recovery undertaking, but also to mix the refuse with sludge and then compost the two in combination. Thus, not only is satisfactory treatment accomplished, but also two wastes that are fast becoming the sources of almost insuperable problems are transformed into a useful product.

## REFERENCES


1. **Golueke, C. G.,** *Biological Reclamation of Solid Wastes,* J. G. Press, Emmaus, Pa., 1977, 18049.
2. **Golueke, C. G. and McGauhey, P. H.,** Reclamation of Municipal Refuse by Composting, Tech. Bull. 9, Sanitary Engineering Research Laboratory, University of California, Berkeley, June 1953.
3. **Epstein, E., Willson, G. B., Burge, W. D., Mullen, D. C., and Enkiri, N. K.,** A forced aeration system for composting wastewater sludge, *J. W.P.C.F.,* 38(4), 688, 1976.
4. **Willson, G. B., Parr, J. F., Epstein, E., Marsh, P. B., Chaney, R. L., Colacicco, D., Burge, W. D., Sikora, L. J., Tester, C. F., and Hornick, S.,** Manual for Composting Sewage Sludge by the Beltsville Aerated-Pile Method, EPA-600/8-80-022, Office of Research and Development, U.S. EPA, Cincinnati, Ohio, May 1980.
5. **Schulze, K. F.,** Rate of oxygen consumption and respiratory quotients during the aerobic decomposition of synthetic garbage, *Compost Sci.,* 1, 36, 1960.
6. **Chrometzka, P.,** Determination of Oxygen Requirements of Maturing Compost, Intern. Res. Group on Refuse Disposal, Inf. Bull. 33, August 1968.
7. **Jeris, J. S. and Regan, R. W.,** Controlling environmental parameters for optimum composting, *Compost Sci.,* 14(2), 8, 1973.
8. **Cardenas, R. R., Jr.,** Rapid moisture determination in compost for process control, *Compost Sci.,* 18(1), 14, 1977.
9. **Niese, G.,** Experiments to Determine the Degree of Decomposition of Refuse by Itself-Heating Capability, Intern. Res. Group on Refuse Disposal, Inf. Bull. 17, May 1963.
10. **Rolle, G. and Organic, E.,** A New Method for Determining Decomposable and Resistant Organic Matter in Refuse and Refuse Compost, Intern. Res. Group on Refuse Disposal, Inf. Bull. 21, August 1964.
11. **Moller, F.,** Oxidation-Reduction Potential and Hygienic State of Compost from Urban Refuse, Intern. Res. Group on Refuse Disposal, Bull. 32, August 1968.
12. **Schulze, K. F.,** Relationship between mosture content and activity of finished compost, *Compost Sci.,* 2, 32, 1964.
13. **Obrist, W.,** Enzymatic Activity and Degradation of Matter in Refuse Digestion: Suggested New Method for Microbiological Determination of Degree of Digestion, Intern. Res. Group on Refuse Disposal, Inf. Bull. 24, September 1965.
14. **Lossin, R. D.,** Compost studies, *Compost Sci.,* 11, 16, 1970.
15. **Haug, R. T.,** Engineering principles of sludge composting, *J. Water Pollution Control Fed.,* 51, 2189, 1979.
16. **Haug, R. T.** *Compost Engineering Principles and Practice,* Ann Arbor Science Publishers/The Butterworth Group, Ann Arbor, Mich. 1980.
17. **Finstein, M. S.,** Composting microbial ecosystem: implications for process design and control, *Compost Sci./Land Util.,* 21, 25, 1980.
18. **Halley, E. M., Jr., Shea, T. G., and Braswell, J. W.,** Process dynamics in aerated static pile composting, paper presented at the 53rd Annu. Water Pollut. Control Fed. Conf., Las Vegas, Nev., September 29 to October 3, 1980.
19. Composting of Municipal Solid Wastes in the United States, U.S. EPA SW-47, 1971.
20. **Stone, G. E., Wiles, C. C., and Clemons, C. A.,** Final report on Joint U.S. EPA-TVA Composting Project with Operational Data, 1967 to 1971, Vol. 1 and 2, EPA/530/SW-31r.2, U.S. EPA, November 1975.
21. Economic indicators, *Chem. Eng.,* 86(23), 7, 1979.
22. Terex 74-51 Composter, brochure, Form No. S-7050, Terex Division of G. M., Hudson, Ohio, April 1973.
23. **Harding, C. I.,** Recycling and utilization, *Compost Sci.,* 9(1), 4, 1968.
24. **Spitzer, E. F.,** Composting works in Houston, *Am. City,* 82, 97, 1977.

25. Anon., $1 to 1/2 million composting plant closes, *Refuse Removal J. Solid Wastes Manage.*, 11, 23, 1968.
26. **Gotaas, H. B.,** Composting, W. H. O. Monogr. Ser. No. 31, World Health Organization, Geneva, 1956.
27. **Diaz, L. F., Trezek, G. T., and Savage, G. M.,** Market Potential of Materials and Energy Recovered from Bay Area Solid Wastes, report prepared by the University of California for the Calif. State Solid Waste Manage. Board, College of Engineering, University of California, Berkeley, March 1976.
28. **Diaz, L. F., Cooper, R. C., Scarpace, L. P., Trezek, G. J., and Golueke, C. G.,** Public Health Aspects of Composting Combined Refuse and Sludge and of the Leachates Therefrom, final report for the State of Calif. Solid Waste Manage. Board, University of California, Berkeley, June 1977.
29. **Golueke, C. G., Lafrenz, D., Chaser, B., and Diaz, L. F.,** Composting combined refuse and sewage sludge, *Compost Sci.*, 21, 42, 1980.

# Chapter 7

# ENVIRONMENTAL ASPECTS

## I. INTRODUCTION

Typically, a recovery system affects two distinct, yet related, environments, namely, the environment within the resource facility itself and that external to the facility. The first environment is the one in immediate contact with individuals employed in the facility. Its health-related effects are essentially limited to facility employees, and are generally discussed under the heading "Occupational Health and Safety". The second environment is the more all-inclusive, since it includes the population at large. Hence, factors affecting it have a correspondingly broader impact. Generally, the occupational environment is more amenable to control than is the external environment. Therefore, engineering and administrative controls are more easily applied to the occupational health and safety hazards existing within a facility. Consequently, of the two, the general environmental health risks may require the more rigorous monitoring system.

## II. EXTERNAL ENVIRONMENT

The external environment may be considered in terms of two aspects, one of which relates to standard of living or environmental quality and the other to the public health. However, it is difficult to determine at which point deterioration of environmental quality begins to have an adverse impact on public health. Perhaps, the two are inextricably bound together. Nevertheless, for the sake of convenience the two are treated as individual entities in this chapter.

Environmental quality is perhaps a better term than "standard of living", since the latter includes wage level, material possessions, and access to modern conveniences. The reference in this section is to pleasant as contrasted to unpleasant or satisfactory vs. unsatisfactory to the populace at large. Factors that determine whether or not an environment is satisfactory can be considered in terms of aesthetics and of irritability. A sharp line of demarcation does not exist between the two groups because an aesthetically objectionable environment rapidly becomes a source of irritation. Factors other than aesthetics and irritability are sound climate, population density, landscape, and a host of others. However, as far as resource recovery is concerned, major factors to be considered in determining effect on quality of environment are aesthetics and sound.

### A. Aesthetics

The two major elements that come to mind when considering aesthetics are appearance and odor. Of the two, the easier to manage is appearance. Undoubtedly, the best way to attain an unobjectionable appearance is by enclosure, containment, or concealment. Enclosure can be by way of a well-designed structure. Type of enclosure is determined by a number of factors, not the least of which is economics. Others are inherent unsightliness of the operation, proximity to residential areas, or more broadly, extent of exposure to public view. At times, concealment can be substituted for enclosure by suitable landscaping, as for example, planting shrubbery and low growing trees around the structural components. Of course, all such efforts at beautification can be rendered ineffective through "poor housekeeping". If debris is allowed to accumulate on the premises, the overall impression becomes thoroughly unfavorable.

Offensive odors are more difficult to control than is poor appearance. The problems with odors ultimately stems from the extreme sensitivity of the olfactory nerve of the average individual. Thus, the threshold level for detection of an odor may involve a concentration of the offensive agent as low as a few parts per billion. If the offensive agent is airborne, the boundaries of the affected area may extend far beyond the confines of the facility. Another factor that contributes to the odor problem is the variation between individuals in deciding whether or not an odor is objectionable. Perhaps the safest interpretation in resource recovery (and in waste treatment in general) is that *any* odor is objectionable.

Odor control must be maintained from the time the wastes are brought to the plant until the finished product is discharged. A safe course to follow is to enclose any malodorous process and then ventilate the enclosure such that the discharge air stream can either be elevated to combustion temperature or be passed through a filter or scrubber. Because the incoming waste is per se malodorous and some odor inevitably "escapes" at the time of transfer from truck to the receiving hopper, the property should be surrounded by a suitably broad buffer zone.

**B. Public Health**

With resource recovery, contact with the product produced in its accomplishment is an important point of interaction between the populace and the environment as modified by the resource recovery activity. The other three points are the air, water, and land.

Sound is an important element in determining the quality of environment. Indeed, there is strong evidence that uncontrolled sound has a definitely deleterious effect on public health. Even though its intensity may not be at a damaging level, any noise of mechanical origin soon becomes objectionable.

*1. Agents*

The agents through which resource recovery may exert an effect on the individual, and hence on the public at large, may be classified into two main groups, biological and nonbiological.

The biological agents are an integral part of the incoming waste stream. They fall into the broad groups of parasites, fungi, bacteria, and viruses, and hence are not "artifacts" of the processing involved in resource recovery. (For convenience, the parasites, fungi, bacteria, and viruses are referred to as "pathogens" or disease-causing organisms in this discussion.) The microbial content of the municipal waste stream received relatively little attention until the early 1970s. While it is true that some work had been done on the subject in the 1950s and early 1960s, the concern was mostly with respect to the effect of composting on the destruction of pathogens, and the analytical procedures left much to be desired.

Among the first to apply "modern" techniques to microbial assays of refuse were Peterson[1] and Cooper et al.[2] According to their findings and judging from reports on the incidence of indicator organisms in leachates from landfills, raw refuse has a microbial population of enteric origin that rivals that of raw sewage sludge. Thus, Cooper et al.[2] report coliform counts in samples of refuse (wet weight) to be as much as $1.3 \times 10^8$/g, of fecal coliforms $5.4 \times 10^7$, and of fecal streptococci $2.48 \times 10^8$. The comparable population sizes by no means imply that refuse ranks with raw sewage sludge with respect to potential pathogenic content. One difference between the two materials, or perhaps to express it more accurately, between the two habitats, is that conditions in a waste-water stream are more conducive to the survival of pathogens than are those in the municipal solid waste stream. A far more important difference is the fact that practically the entire population of enteric indicators found in refuse is of animal "origin", while those in sewage sludge are of human "origin".[2] At the time of this writing, information on the incidence of pathogenic viruses in refuse was insufficient to warrant a general statement on the subject, although work by Diaz et al.[3] indicates that the probability of refuse containing even a small concentration of pathogenic viruses is rather remote. The scarcity of information on viruses is a result of the lack of adequate techniques for isolating the viruses from refuse samples. Present isolation techniques are at best from 20 to 40% effective.

Three principal characteristics can render a nonbiological agent deleterious to the public health: (1) an inherent toxicity and corrosiveness to humans; (2) a physical configuration; and (3) a dimension that renders it respirable by humans. Examples of toxic substances are the heavy metals and carcinogenic organic compounds such as PCB. Acids and gases (e.g., $SO_2$) formed during processing are examples of corrosive agents. Representative of harmful physical configurations and dimensions are particulates contained in air streams discharged into the environment. Particulates constitute a potential hazard only when they are of a size that permits them to be carried into the alveoli of the lungs where they can accumulate. Particles as large as 2 μm are quite likely to reach the alveoli,[4] and those as much as 10 μm will be deposited deep in the respiratory tract.[5] Particulates most likely to be harmful are fibrous in nature. Of the fibers, the asbestos fibers are definitely carcinogenic. Other fibers of public health significance to be found in refuse and which, therefore, can become a discharge pollutant from a resource recovery operation are glass fibers, mineral wool, cellulose fibers, cotton, wool, and plastic synthetic fibers.

#### *2. Modes of Entry into the Environment*

The mode of entry of biological and nonbiological agents into the external environment varies with the nature of the resource recovery system in operation. Thus, contamination of the external environment by thermal recovery systems is principally through air pollution. On the other hand, with composting operations, any contamination would be mostly from the use of the product, although a poorly conducted one could be attended by air pollution in the form of objectionable odors. Because the process determines mode of entry, the various modes are discussed in terms of the principal resource recovery groups.

### C. Air Pollution

Probably the most likely mode of entry of biological and nonbiological agents into the external environment from a resource recovery operation would be by way of dust particles borne in the air stream discharged from the facility. As is shown in the section on occupational health, dust particles and aerosols can transport a sizeable population of microbes. However, the problem is readily minimized by the application of dust suppression measures during waste handling and by processing air discharges through air filters. The effectiveness of air filtration was indicated by results obtained in research conducted at the University of California.[3] In the university test, air vented from the air classifier to the exterior, after having been passed through an air filter, was sampled through conventional sampling techniques. The bacterial and particle counts of the vented air were not significantly different from those of the ambient air.

Following the front-end steps (sorting, shredding, and classification), the effect on ambient air quality varies according to the type of resource recovery being applied, and even more particularly, to the quality of the control imposed on the operation. The two principal potential offenders in terms of impact on air quality occur in thermal processing for energy conversion and in composting.

*1. Thermal Processing*

The point of interaction between the external atmospheric environment and a thermal energy conversion process is the stack. The mode of pollution is the release of the products of combustion into the atmosphere by way of the stack discharge. The formation of combustion products is not restricted to direct conversion processes (e.g., incineration, cofiring of RDF). It also occurs in one of the stages of current pyrolysis systems.

Combustion products generally are divided into two broad classes, namely, particulates and gases. The amounts and characteristics of the two groups are functions of the nature of the material (fuel) being combusted and of the manner of combustion. Despite the heterogeneity of municipal refuse, certain generalizations can be made with a considerable degree of accuracy about the characteristics of typical products.

Particulates (fly ash) are finely divided solid materials entrained in the flue gases emitted from a furnace. The particle size of fly ash ranges from less than 5 to 120 μm or more. Excepting at the 15- to 20-μm sizes, no one particle size seems to predominate. The quantity of fly ash generated during the combustion process depends upon a number of factors, and especially on the rate of input of underfire air. It can be from as little as 4 kg to as much as 25 kg/Mg of refuse burned. The density of the particulates varies from 2.65 to 3.71 g/cm$^3$. The maximum lead content of the particulates is on the order of 78,000 ppm of particulates; the cadmium, 1900 ppm; and the zinc, 140,000 ppm. These concentrations are not great enough to pose any environmental difficulties in combustion, even in the failure to provide air pollution control.[6]

Gaseous combustion products include sulfur dioxide ($SO_2$), hydrogen chloride (HCl), oxides of nitrogen ($NO_x$), and hydrocarbons (HC). Of these gases, $SO_2$ occurs only in trace amounts, the reason being the low sulfur content of refuse. For example, the $SO_2$ emissions from the incineration of refuse ranges from 56 to 195 ppm, whereas, when coal is fired in a utility boiler, the $SO_2$ emission may be as much as 1200 to 2300 ppm.[6,7] Cofiring coal (75 to 90%) and RDF (10 to 25%) results in $SO_2$ emissions of 1000 to 1200 ppm. The rather small reduction in $SO_2$ emissions accomplished by cofiring coal and RDF obviously is due to the fact that RDF constituted only a small percentage of the total mixture.

The extent of the emission of oxides of nitrogen ($NO_x$) from the combustion of RDF probably is more a function of the manner in which the furnace is operated than of the nature of the RDF. Thus, the more excess air that is introduced, the greater is the amount of $NO_x$ that is emitted. On the other hand, reduction in furnace temperatures leads to a lowering of $NO_x$ emissions. The amount of HCl emitted in the combustion of refuse (or RDF) exceeds that in the burning of coal. When refuse is incinerated the HCl emissions may range from 200 to 1250 ppm.[8] If it is pyrolyzed, the emissions from the pyrolytic combustor could be as low as 13 to 37 ppm. On the other hand, when coal alone is fired, the HCl emission is on the order of 14 to 31 ppm. Expressed in terms of kilograms of HCl/Mg of refuse fired, the emissions may range from 1.9 to 4.3 kg/Mg.[9] As with $NO_x$, the amount of hydrocarbons emitted is more a function of the manner in which the furnace is operated than of the fuel burned in it. Reported concentrations are as low as 2 to 30 ppm.[6]

Methods of achieving air pollution control are described in summary fashion in the publication "Municipal-Scale Incinerator Design and Operation" issued by the U.S. Environmental Protection Agency[10] and in somewhat greater detail in the report "Municipal-Scale Thermal Processing of Solid Wastes" prepared by RECON Systems, Inc.[11] The latter report represents an updating of the "Municipal-Scale Incinerator Design and Operation" publication.

Of the gaseous emissions, it is unlikely that the amount of $SO_2$ emitted would be such that any measures would have to be taken to remove it. Nitrogen oxide emissions from waste-to-energy facilities may have to be controlled. Unfortunately, at the time of this writing most methods for controlling $NO_x$ emissions are expensive and subject to a great deal of debate. Hydrogen chloride is an irritant to the eyes and the respiratory tract and can constitute a health hazard if released in large enough amounts during combustion. That such may be the case is becoming more likely because of the increasing amounts of plastics to be found in municipal wastes, especially of polyvinyl chloride. Fortunately, the HCl can be readily removed from the stack gases because it is highly soluble in water.

Various types of systems are available for collecting particulates. They include settling chambers, wetted baffle or baffle spray system, cyclones and multiple cyclones, wet scrubbers, electrostatic precipitators, and fabric filters. Of this list, it is only with the last three named that current and future emission standards can be met. A description of the three is beyond the scope of this book. However, principles as well as capacities and limitations of the three types are ably presented in the report prepared by RECON Systems.[11]

*2. Composting*

In 1978 attention began to be focused on the pronounced output of the spores of *Aspergillus fumigatus* that accompanies the composting of sewage sludge, especially when wood is the bulking material. *A. fumigatus* is regarded as being a pathogen of very low infectivity, and thus is of consequence only to individuals receiving intensive cortisone treatment or recovering from tuberculosis or to the aged or to those in poor health in general. The existence of a hazard, if any, is as yet hotly debated, although the trend is to downplay the danger. Thus far, no record exists of any employee of a compost plant or of anyone living in the proximity of a compost operation having developed Aspergillosis. Nevertheless, as a safety measure, an open-air compost operation should not be sited in close proximity to human habitation.

**D. Water Pollution**

The likelihood of the entry of pollutants into the external aqueous environment is a function of the nature of the particular resource recovery system being used and of the measures taken to treat waste-water discharges from the operation. Aqueous transport becomes important in front-end systems that involve hydraulic processing, such as does the *Black Clawson Hydrasposal®* systems and modifications thereof.[12] Water not recycled within the system must be subjected to conventional waste-water treatment before being discharged into the environment. The microbial and nonbiological contents of the process water undoubtedly matches those of leachate from freshly laid landfills.

With systems that involve the use of water at some point in the back-end phase of the operation, the degree of contamination is a function of the nature and condition of the raw material being processed and the nature of the process. For example, in the recovery of paper fiber for reuse in paper manufacture, as it is done in the system developed by Cal Recovery Systems, Inc., water is used in the pulping of the waste paper and in the subsequent cleansing of the fibers.[13,14] The contamination of the process water in the Cal Recovery Systems process is far less than that with a front-end hydraulic system. For example, with the former, the BOD is on the order of 300 to 400 mg/ℓ, whereas with the latter it is as high as 2000 to 6000 mg/ℓ. Inasmuch as the relative chances and extent of microbial contamination parallels the extent of increase in BOD, there is some possibility, albeit remote, that pathogens may be present in the discharge process waters.

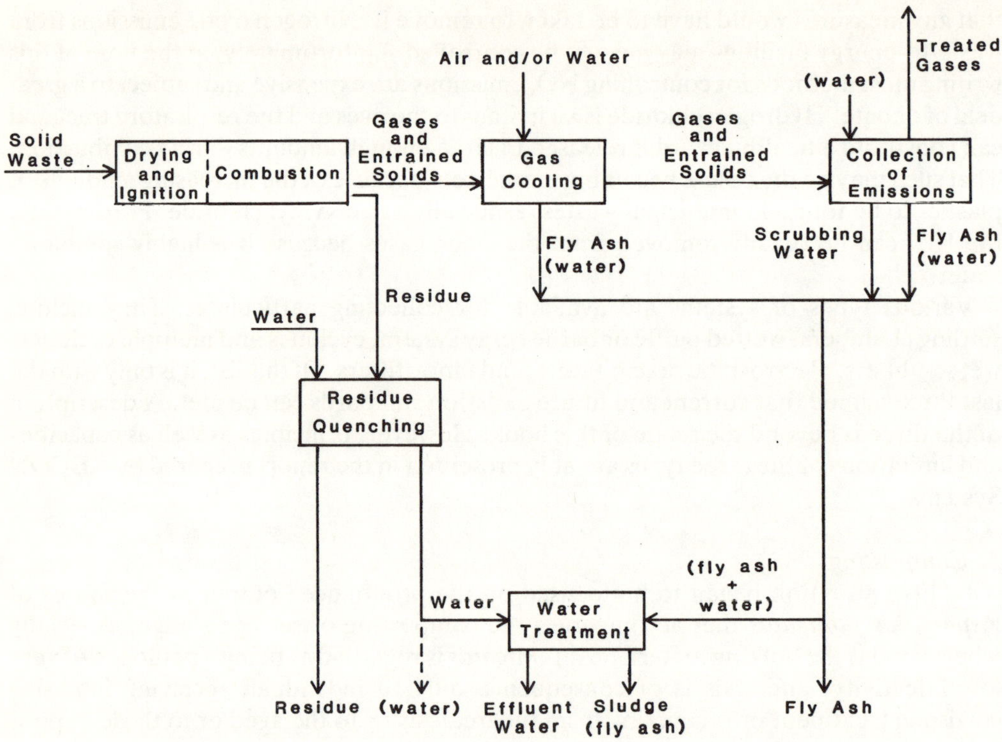


FIGURE 1. Points of water pollution in incineration.

In thermal systems, major sources of water pollution are in air-pollution control and in quenching the incinerator residue (ash) prior to disposing of it. The sources and their position in the overall incineration process are shown in the flow diagram in Figure 1. An idea of the chemical quality of the process waters may be gained from the information in Table 1. Obviously, it requires considerable treatment before it can be discharged into the environment, and less if it is recycled within the plant. Usually, passage through a settling chamber upgrades the process water to a level at which it can be recycled within the plant.

Unless each and every particle of waste is combusted in the incinerator, viable microbes can be found in the incinerator residue, i.e., microbes that survived passage through the furnace. Furthermore, unless the incinerator is properly operated, microbes will be found in the quench water. For example, Peterson[1] found the coliform density in quench waters from two well-operated incinerators to be less than 100 organisms per 100 mℓ, whereas with two unsatisfactorily operated units, she obtained counts as high as $1.7 \times 10^4$ to $2.9 \times 10^4$ organisms per 100 mℓ.

## E. Use of the Product

Two broad classes of reclaimed products warrant attention, namely, paper reclaimed for the fashioning into new items and the use of compost in horticulture and crop production. The public health aspects in the fashioning of new products is in the use of the new items. Generally, the constraints apply only when the reclaimed fibers are to be used in food packaging. Because of the detection of microbial contaminants in paper products manufactured from fiber reclaimed from municipal wastes, the FDA forbids the use of such material in the manufacture of food containers and wrappings. The

## Table 1
## CHARACTERISTICS OF INCINERATOR WASTE WATER

| Characteristic | Plant 1[a] residue quench | | | Plant 2[b] residue quench | | | Plant 3[c] fly ash effluent | | |
|---|---|---|---|---|---|---|---|---|---|
| | Max | Min | Av | Max | Min | Av | Max | Min | Av |
| pH | 11.6 | 8.5 | 10.4 | 11.8 | 9.4 | 11.0 | 6.5 | 4.8 | 5.8 |
| Dissolved solids, mg/ℓ | 9005 | 597 | 3116 | 7929 | 901 | 3894 | 9364 | 7818 | 8838 |
| Suspended solids, mg/ℓ | 2680 | 40 | 671 | 1888 | 72 | 653 | 398 | 208 | 325 |
| Total solids, (% volatile) | 53.6 | 18.5 | 36.3 | 47.4 | 7.8 | 26.6 | — | — | — |
| Hardness ($CaCO_3$)mg/ℓ | 1574 | 216 | 752 | 1438 | 174 | 904 | 2780 | 2440 | 2632 |
| Sulfate ($SO_4$)mg/ℓ | 430 | 110 | 242 | 565 | 115 | 300 | 1350 | 1125 | 1250 |
| Phosphate (PO)mg/ℓ | 55.0 | 0.0 | 23.3 | 225 | 2.5 | 33.8 | 15.0 | 11.5 | 13.0 |
| Chloride (Cl)mg/ℓ | 3650 | 50 | 627 | 1940 | 128 | 868 | 3821 | 3077 | 3543 |
| Alkalinity ($CaCO_3$)mg/ℓ | 1250 | 215 | 516 | 1290 | 337 | 682 | 28 | 16 | 23 |
| 5-Day BOD at 20 C | — | — | — | — | — | — | 13.5 | 6.2 | 8.8 |

[a] Plant 1—110 TPD, batch, residue quench only.
[b] Plant 2—500 TPD, batch, residue, quench only.
[c] Plant 3—300 TPD, continuous feed, fly ash effluent only.

Modified from DeMarco, J., Keller, D. J., Leckman, J., and Newton, J. L., Municipal-Scale Incinerator Design and Operation, U.S. Department of Health, Education, and Welfare, P.H.S. (presently U.S. EPA), 1969.

carry-over is not limited to microbes. Apparently, malodorous substances are also carried through the reclamation and manufacture processes.

The situation with the use of the compost product is more complex. The impact on the public health hinges upon the use that is made of the compost and, of course, upon the pathogen and toxic substance concentration of the product.

The public health aspects differ somewhat when inanimate contaminants are considered. With certain nonbiological contaminants, an unfavòrable impact can be exerted even at a significant distance from the site of application. Heavy metals and toxic materials most likely would not pose a problem if only refuse, agricultural, and garden debris are composted. They can become a problem when refuse and sewage sludge are composted together. In the latter case, the sludge would be the main contributor. A very possible route of contamination is the leaching of soluble and solubilized contaminants into the ground water or transport to a nearby open body of water. The chemical and organic content of leachate from compost is a function of the nature of the raw materials and of the extent to which the material was composted. Diaz and Trezek[15] demonstrated a direct relation between the age of the compost and the concentration of potential pollutants in the compost product. They found that the BOD, COD, and the concentrations of total and volatile solids and of volatile acids declined at a rate proportionate to the age of the compost. The situation with metals was not as clear-cut. While the concentrations of Na, P, and Fe declined with the passage of time, the reverse proved to be true with those of Cu, Pb, and Cd. Perhaps the latter remained insoluble, and hence could not be transported. Age of compost had no apparent effect on concentrations of Cr, Ni, and Zn.

Another potential route of contamination is that of nitrogen leaching to the ground water. Such an event would occur only if the land were loaded with compost to the extent that more nitrogen is released than can be assimilated by the crop. Excess nitrogen is oxidized by soil bacteria to a soluble form, namely $NO_3$. If it is above a certain level, nitrate in drinking water can be harmful to the health of one imbibing it.

Chances of an unfavorable impact in terms of pathogens are only marginal when the compost is used in land reclamation, especially at seldomly visited sites (e.g., reclaimed strip mines). The chances become greater when the product is used in urban recreational areas, providing, of course, the compost contains pathogens. Finally, chances of an unfavorable impact are greatest when the product is used in the production of root crops of which the edible portion is eaten either raw or unpeeled.

The presence of pathogens in the compost product is determined by a number of factors, among which the type of the raw material, level of temperature reached in the pile and persistence of that level, and detention period imposed in the process are especially important. Certainly, if the raw material includes human excrement (raw sewage sludge), the possibility of pathogenic organisms being present is accentuated. On the other hand, if only plant debris is composted, chances are exceedingly remote that pathogens would occur.

Temperature is an exceedingly important factor in determining pathogen survival. Unfortunately, a trend exists in which too much confidence is placed on the bactericidal effectiveness of the elevated temperatures encountered in composting. While it is true that the temperatures substantially exceed the thermal death point of practically all pathogens, it is a fact that not all parts of the pile are necessarily exposed to the high temperatures at any one time. Consequently, some pathogens may escape destruction and later contaminate the material that had been exposed to the high temperatures. Regardless of compost method, it would be rash to expect that all material in a pile exceeding several tons in weight would be simultaneously exposed to identical conditions. Another factor that could contribute to the uncertainty is the moisture content of the pile. Moist heat is far more lethal to pathogens than is dry heat.

A factor that can compensate for the inadequacy of high temperature is time. Although time is not a factor, strictly speaking, it functions as one. If enough time is allowed, factors other than temperature come into play. These factors are destruction of nutrient source, competition with indigenous populations, and antibiosis. Pathogens generally are rather specific in their nutrient requirements. During the compost process, the nutrients are decomposed, and thus, the pathogens lose a source of nutrients. Competition is very important. Generally, indigenous microbial populations, being well adapted to the environment in the wastes, are in an excellent condition to compete with the introduced (pathogenic) populations. Antibiosis is a very devastating factor for susceptible pathogens. The principle source of the antibiotics are the fungi and actinomycetes invariably found in a compost pile. If the raw material included raw sludge, a year-long interval between completion of composting and use of the product in food production would not be excessive.

## III. OCCUPATIONAL HEALTH AND SAFETY

### A. Introduction

To explore the occupational health and safety factors in resource recovery in all its aspects, i.e., both the preparation of the wastes for reuse ("front-end" plant) and actual making use of the reclaimed resource ("back-end" plant) is beyond the scope of this book. Hence, the emphasis herein is on the "front-end" plant because it is a feature common to all resource recovery operations and has many novel features not shared in conventional industrial practice.

The occupational health hazard may be treated under two broad headings, namely, health and safety. Potential health hazards (dust and particulates) are principally in three forms: air pollution, physical contact with the wastes stream, and excessive noise. The safety aspects relate to the heavy machinery used in a centralized processing plant.

**B. Health**

*1. Physical Contact*

Although the potential is there, this aspect thus far has been associated with a very low profile in terms of being a factor in the incidence of disease among workers in resource recovery operations. Undoubtedly, a decisive factor in the low profile is the practice of wearing gloves while working in the plant. In addition to requiring the workers to wear gloves, insisting that employees rigidly adhere to basic hygienic practices will result in the minimizing of chances of infection by the manual-oral route. The provision of suitable sanitation facilities for the work force goes far to ensure such a practice. As a part of the facility, the usual first-aid supplies are a must. For example, disinfectants and Band-aids® should be readily accessible. Outer clothing worn in the plant should not be worn outside the plant, e.g., at home, unless care is taken to keep it separate from other clothing.

Finally, all employees should be immunized against tetanus, and their physical condition should be carefully monitored. Being in good health is the best protection against disease-causing organisms that may occur in refuse.

*2. Noise*

The importance of noise with respect to individual well-being is becoming widely recognized. The most easily observed of the adverse health effects of exposure to excessively high noise levels is loss of hearing. Excessive noise wears out the nerve cells of the inner ear, and if exposure is of a long term, the cells are destroyed. The result is a permanent impairment or even complete loss of hearing. Although no definite answer exists as to the level at which continuous noise becomes dangerous to the hearing, the consensus places the level at 80 dB. As of 1979, the allowable industrial noise level in the U.S. was 90 dB for 8 continuous hr. It has been estimated that if that level were allowed to stand, about one fifth of the work force (all industry) would eventually suffer disabling loss of hearing.[16]

In establishing its standard for noise, the Occupational Safety and Health Administration (OSHA) ties noise level with continuity and frequency of exposure. Thus, in 1976, the top limit for momentary impact noise was 140 dB. If such levels are reached at intervals of less than 1 sec, the noise is to be regarded as being continuous. Allowable noise levels as functions of time intervals are plotted in Figure 2.

Hearing loss is not the only negative effect that workers suffer from noise. Noise creates stress in the individual and thereby leads to the many unfavorable physiological conditions that result from stress. For example, the incidence of heart disease is greater among workers in noisy industries. Moreover, researchers report a greater incidence of high blood pressure among workers exposed to high levels of noise.

Safety and noise are related in that at excessively high levels of noise, warning signals may be masked. As a result, the imminence of a dangerous situation might not be anticipated in time for counter measures to be taken.

Workers and companies are poles apart as to the best approach to the noise problem. Most companies propose the use of hearing protectors, simply because they are inexpensive and the burden of noise control is placed upon the workers. However, sound reasons exist for making the use of personal hearing protectors a last resort. Ear plugs easily work loose and frequently cause infections. Moreover, they can mask warning shouts and signals. The best approach is to treat hearing protectors as temporary measures. For the long term, recourse must be had to alterations in operational and equipment design to reduce noise.

As of this writing, reports on noise levels encountered in resource recovery facilities are rare. One of the earlier studies was conducted (1976) at a pilot plant located at the

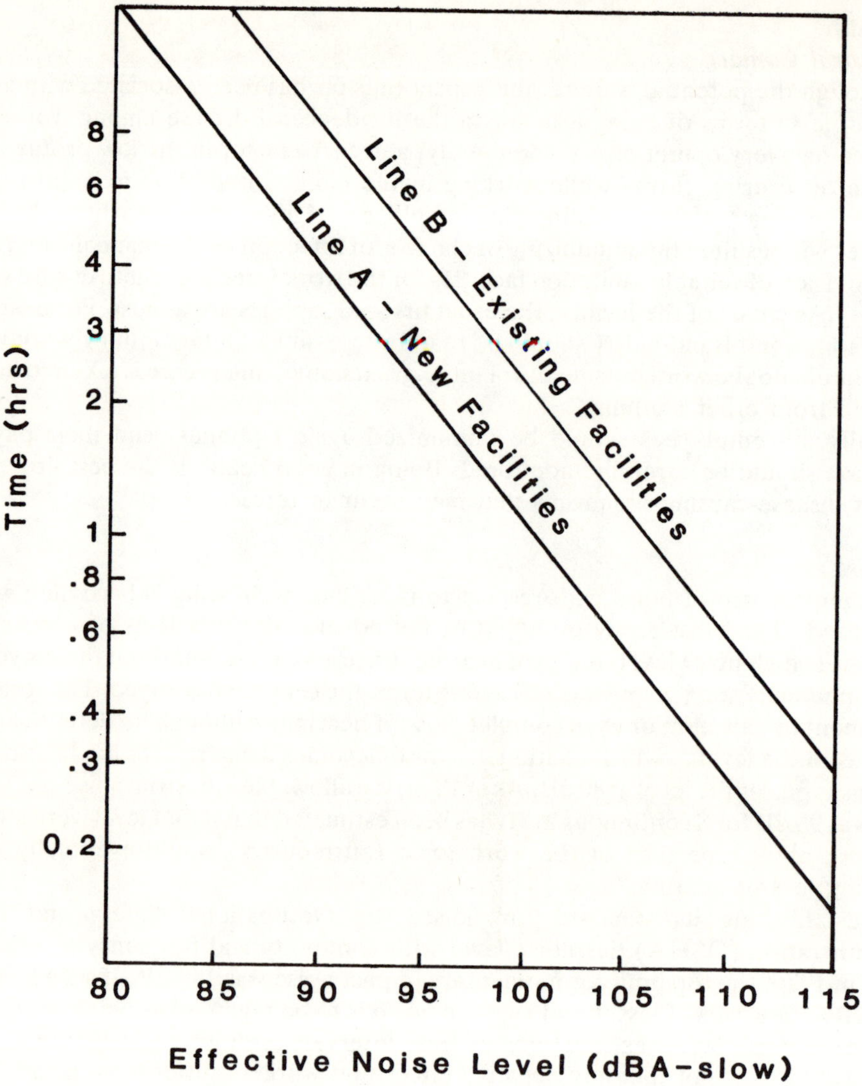


FIGURE 2. Allowable noise as a function of duration.

Richmond facility of the University of California.[17] Inasmuch as the university's pilot plant has a throughput capacity of 4 tons/hr directly from the collection truck, and its shredder is a full-scale model (10 ton/hr), observations made on noises generated in its operation may be considered to be applicable to the waste recovery industry in general. The layout of the university plant and the points at which observations were made are illustrated in Figure 3. A type-1 sound level meter placed at a height of 1.1 m was used to determine effective noise levels at the points indicated in the figure. Results obtained in the study are listed in Table 2 and are fairly typical of the noise levels to be encountered in general practice, although some specialists may feel the levels are somewhat low. According to the table, the peak levels were on the order of 93 to 96 dB-slow, and they were observed in the immediate vicinity of the shredder. These levels were reached on the order of four or five times per hr and are well below the maximum of 140 dB for momentary impact noise allowed by OSHA. According to the curves in Figure 2, workers in the vicinity of the shredder and air classifier (A, B, and C in Figure 3) for

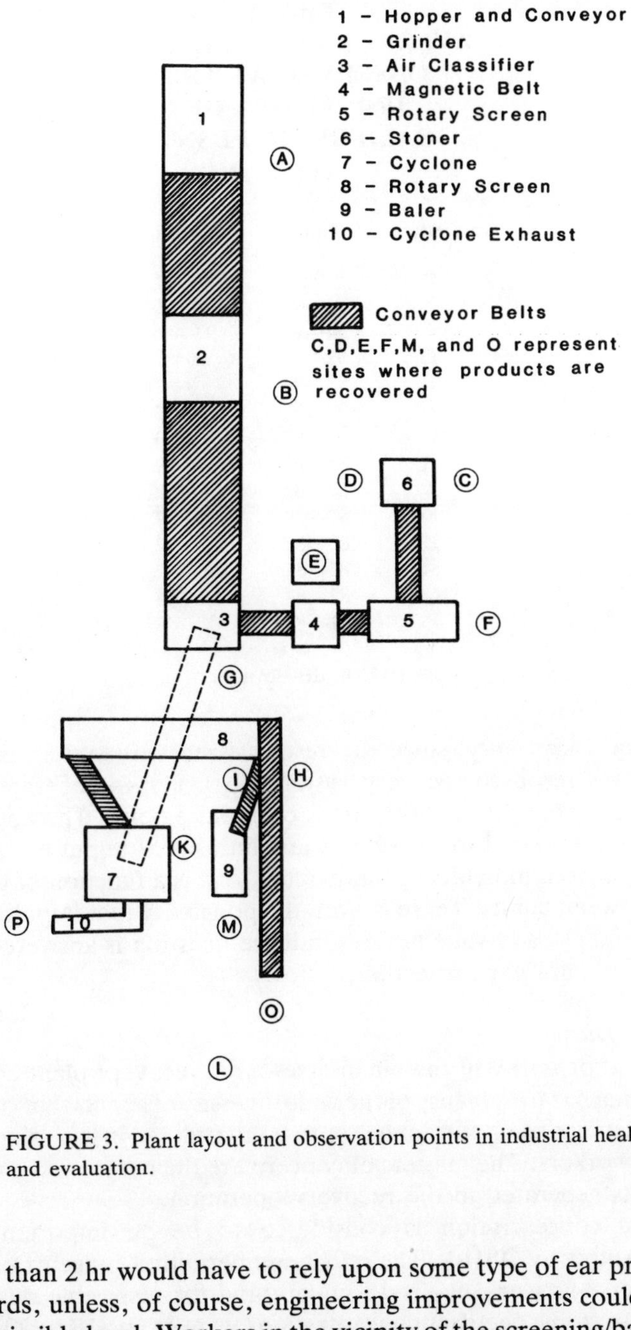


FIGURE 3. Plant layout and observation points in industrial health and evaluation.

periods longer than 2 hr would have to rely upon some type of ear protection to meet OSHA standards, unless, of course, engineering improvements could bring the noise down to a permissible level. Workers in the vicinity of the screening/baling operation (I and K in Figure 3) could work for almost 5 hr without ear protection.

The study also included an octave band analysis made by means of type II octave band analysis[18] at each observation point to determine whether or not any particular frequency had a very high noise level. Since no single band was found to have an inordinately high noise level, it may be concluded that no particularly damaging noise levels at any one frequency was masked by the procedure of measuring levels over the entire spectrum.

The preceding analysis of damage risk from noise is based on National Institute of Occupational Safety and Health standards for continuous noise. This fact introduces a

**Table 2**
**AVERAGE NOISE LEVELS[a] OBSERVED AT THE UNIVERSITY OF CALIFORNIA PLANT (1975)**

| Site | Test 1 | Test 2 |
|---|---|---|
| A | 94 | 93—96 |
| B | 94—95 | 95—96 |
| C | 94—96 | 93—96 |
| D | 92—93 | 94 |
| E | 93—94 | 94 |
| F | 90—96 | 91—96 |
| G | 91—93 | 91—93 |
| H | 86—88 | 87—89 |
| I | 87—89 | 87—89 |
| K | 87—89 | 88—89 |
| L | 80—81 | 81—82 |
| M | 85—87 | 86—87 |
| O | 83—84 | 83—84 |
| P | 81—82 | 81—82 |

[a] Noise level in dB-slow.

certain amount of uncertainty, since the resource recovery system observed in the study—and perhaps resource recovery systems in general—generates a noise that actually is a series of continuing short bursts of impulse noise. The uncertainty stems from the lack of data related to the effects and hazards of impulse noise. The effect undoubtedly varies from individual to individual, and is a function of the individual's ear to recover between bursts. There is even the possibility that impulse noise may be more hazardous than steady-state noise. Until the question is answered, the workers should be provided with ear protection.

*3. Air Pollution (Dusts)*

Dusts suspended or borne in the air in a resource recovery plant can constitute a unique occupational hazard because of the wide diversity of agents that can be involved. Consequently, dusts are an important factor to be considered in assessing the impact on the health of the workers. The matters of concern are the nature and concentration of the airborne dusts generated in the recovery operation.

The total dust concentration is considered to be an important measure of environmental hazards by OSHA. The maximum permitted concentration of inept or nuisance dusts per 8-hr work day is 15 mg/m$^3$, and for respirable dusts, 5 mg/m$^3$.

In the California study, the dust concentrations were from 118 to 202 mg/m$^3$ in the vicinity of the shredder, and not much lower in the vicinity near the screens (I in Figure 3). Consequently, with the poor dust control obtained in the university study, workers would have to wear face masks equipped to filter out dust. However, in a study conducted by Duckett[19] at the National Center for Resource Recovery's (NCRR) Equipment Test and Evaluation facility, the greatest dust level attained was only about 38 mg/m$^3$. The typical level probably is somewhere between the two extremes. An attempt to measure the particle size distribution was made with the use of six-stage Anderson sampler.[20] Particles were impacted on six nutrient agar plates, one at each stage. Results obtained in the attempt are listed in Table 3. The plate counts on the

**Table 3**
**PARTICLE COUNT AND DISTRIBUTION PER CUBIC FEET OF AIR**

| Stage[a] | Size (μm) | Ambient[b] | | Operating[c] | |
|---|---|---|---|---|---|
| I | 8 | 5 (20) | 8 (21) | 52 (16) | 101 (16) |
| II | 5—10.5 | 2 (8) | 8 (21) | 50 (15) | 61 (10) |
| III | 3—6 | 2 (8) | 3 (8) | 23 (8) | 64 (10) |
| IV | 2—3.5 | 4 (16) | 7 (18) | 84 (25) | 150 (23) |
| V | 1—2 | 5 (20) | 8 (21) | 72 (22) | 168 (26) |
| VI | 1 | 5 (20) | 4 (11) | 51 (15) | 99 (15) |
| Total | | 25 (100) | 38 (100) | 332 (100) | 643 (100) |

[a] Stage of a six-stage Anderson sampler.
[b] Ambient—sampled on days during which the facility was not in operation.
[c] Numbers in parentheses denote percentages.

various stages do not represent a one-to-one correspondence-to-particle impaction. However, many of the organisms impacted can be assumed to have been associated with a larger item of particulate matter. Consequently, by comparing the numbers of viable particles impacted on each stage during the two conditions (background and operating), some indication of the lung penetration potential of the recovery system can be obtained.

***a. Dust Analysis***

Factors other than size that determine the occupational aspects of dusts are the nature (i.e., composition) of the dust particles and their biological burden. The biological burden is the collection of parasite-resistant forms, pathogens, and pathogenic viruses carried by the dust particles.

The fibrous fraction of dust can have a health significance. A fiber that has justly received considerable attention is the asbestos fiber. Its role as a carcinogenic agent has been well publicized. The few tests conducted up to 1978 on dusts collected in refuse processing plants indicate that the asbestos content is very small.[17,19] Obviously, the content varies from plant to plant, and within a plant, from day to day. However, it can be assumed that excepting under unusual circumstances, the asbestos content of dust in a typical plant will be far below the Threshold Limit Value (TLV).

Fibers other than asbestos to be found in refuse and which would have an industrial health significance are glass fibers, mineral wool, cellulose fibers, cotton, wool, and plastic synthetic fibers. To be significant, the fibers would have to be present as particles of respirable size. The extent, if any, of the hazard from such particles remains to be explored.

The dust may include heavy metals of health significance. Whether or not this is true with dusts generated in a processing plant remains to be demonstrated. Evidence collected thus far is inconclusive.[19]

***b. Microorganisms***

Biological contaminants in the air are in the form of aerosols. The microorganisms in the aerosols may occur as single units, as clumps of organisms, or as adhering to dust particles. Two types of infections involving airborne contaminants are of concern in a resource recovery plant. Both are acquired by way of the respiratory tract, but differ as to the part of the body that may be affected. Thus, one type is limited to the respiratory tract, while the second may affect some other part of the body, e.g., gastroenteric tract.[21]

Reliable information on relationships between solid waste processing and transmission of infectious diseases is scarce or even nonexistent. Epidemiological studies have not revealed any significant differences between sanitation workers and control groups for rates of skin infections or gastrointestinal disorders.[22,23] As yet, experience in terms of time and amount is insufficient to warrant definite conclusions as to an increase, if any, in the incidence of respiratory or other infections due to employment in a resource recovery facility.

The failure to demonstrate a connection between exposure to the dusts in a plant and the incidence of disease does not justify a conclusion that no hazard exists. There is no gainsaying the fact that the bacterial count of the air in a recovery plant is far higher than that of the external environment both in terms of total bacteria and of fecal coliforms and streptococci.[17] Thus, while normal outdoor bacterial levels are seldomly found to be more than 1 or 2 colonies per liter of air, those inside the University of California resource recovery plant ranged from 9 to 13 colonies per liter. The occurrence of coliforms in outdoor air ranges from zero concentration to only a few per cubic meter. On the other hand, inside the university's facility total coliforms ranged from 2200 to $4000/m^3$, and fecal streptococci from 13,800 to $18,900/m^3$. The heavy concentrations of enteric indicators (coliforms and fecal streptococci) isolated from dust samples reflect the heavy concentrations to be found in raw refuse. However, as a result of their studies, Cooper et al.[24,25] conclude that the coliforms and other enteric indicators are mostly of animal origin than of human origin. The high numbers of indicator organisms do furnish sufficient reason for reducing the amount of dust respired by workers employed in a recovery plant, even though the indicators may be of nonhuman origin.

A major reason for the element of uncertainty that pervades all reports on the characterization of the interior environment of recovery plants and of emissions from them is the slowness in the development of standardized sampling and analytical techniques. A good idea of the problems involved and of the progress made in solving them may be gained from the publications by Diaz et al.[17] and by Duckett,[19] and especially from a draft report prepared by Gorman et al.[26]

### C. Safety

Precautions to be taken in a recovery plant are those to be followed when working around any heavy machinery, and are duly listed and described in OSHA regulations and guidelines. A major addition to the usual hazards attending the use of heavy machinery is the danger of an explosion occurring in the shredder as was mentioned in another section. Another important hazard that can develop in the shredding operation is the high-velocity ejection of hard objects (rocks, metal) that may result from the impact of the hammers upon the objects.

# REFERENCES


1. **Peterson, M. L.,** Pathogens Associated with Solid Waste Processing, Progress Report (SW-49r), U.S. EPA, 1971.
2. **Cooper, R. C., Klein, S. A., Leong, C. J., Potter, J. L., and Golueke, C. G.,** Effect of Disposable Diapers on the Composition of Leachate from a Landfill, final report, SERL Report No. 74-3, Sanitary Engineering Research Laboratory, University of California, Berkeley, February 1974.
3. **Diaz, L. F., Cooper, R. C., Scarpace, L. P., Trezek, G. J., and Golueke, C. G.,** Public Health Aspects of Composting Combined Refuse Sludge and of the Leachate Therefrom, final report to state of California Solid Waste Management Board, College of Engineering, University of California, Berkeley, June 1977.
4. **Comroe, J.,** The lung, *Sci. Am.,* 214(10), 56, 1966.
5. **Hatch, T.,** Distribution and deposition of inhaled particles in the respiratory tract, *Bacteriol. Rev.,* 25, 237, 1961.
6. **Duckett, J.,** Health aspects of resource recovery. II. Air pollution, *N.C.R.R. Bull.,* 8(4), 105, 1978.
7. **Kaiser, E. R.,** The sulfur balance of incinerators, *A.P.C.A. J.,* 18(3), 171, 1968.
8. **Kaiser, E. R., and Carotti, A. A.,** Municipal Incineration of Refuse with 2 Percent and 4 Percent Additions of Four Plastics: Polyethylene, Polyurethane, Polystyrene, and Polyvinyl Chloride, in Proc. 1972 Nat. Incinerator Conf., A.S.M.E., New York, June 4 to 7, 1972.
9. **Carotti, A. and Smith, R. A.,** Gaseous Emissions from Municipal Incinerators, (SW-18c), U.S. EPA, 1974.
10. **De Marco, J., Keller, D. J., Leckman, J., and Newton, J. L.,** Municipal-Scale Incinerator Design and Operation, U.S. Department of Health, Education, and Welfare, P. H. S. (presently U.S. EPA), 1969.
11. **Weinstein, N.,** Municipal-Scale Thermal Processing of Solid Wastes, Final Report (SW-133c), U.S. EPA, 1977.
12. **Herbert, W. and Flower, W. A.,** Waste processing complex emphasizes recycling, *Publ. Works,* 102(6), 78, 1971.
13. **Savage, G. M., Diaz, L. F., and Trezek, G. J.,** Fiber Recovery from Municipal Solid Wastes, in Proc. 6th Mineral Waste Utilization Symp., U.S. Bureau of Mines, Chicago, Ill., May 2 and 3, 1979, 222.
14. **Trezek, G. J. and Golueke, C. G.,** Availability of cellulosic wastes for chemical or biochemical processing, *AIChE Symp. Ser.,* 72(158), 52, 1976.
15. **Diaz, L. F. and Trezek, G. J.,** Chemical characteristics of leachate from refuse-sludge compost, *Compost Sci./L and Utili.,* 20(3), 27, 1979.
16. **Stansbury, J.,** Noise in the workplace, *E.P.A. J.,* 5(9), 18, 1979.
17. **Diaz, L. F., Riley, L., Savage, G., and Trezek, G. J.,** Health aspect considerations associated with resource recovery, *Compost Sci.,* 17(3), 18, 1976.
18. **Peterson, A. P. and Gross, E. E., Jr.,** *Handbook of Noise Measurement,* General Radio, Concord, Mass., 1974.
19. **Duckett, E. J.,** Health aspects of resource recovery. I. Occupational health and safety, *N.C.R.R. Bull.,* 8(3), 62, 1978.
20. **Anderson, A. A.,** A new sampler for the collection, sizing, and enumeration of viable airborne particles, *J. Bacteriol.,* 176, 471, 1958.
21. **Pappagianis, D.,** Some characteristics of respiratory infection in man, in *An Introduction to Experimental Aerobiology,* Oimmick, R. L. and Akers, A. B., Eds., John Wiley & Sons, New York, 1969.
22. **Sliepcevich, E. M.,** Effect of Work Conditions upon the Health of the Uniformed Sanitationmen of New York City, Ph.D. dissertation, New York University, New York, 1955.
23. **Cimino, J. A.,** Health and safety in the solid waste industry, *Am. J. Publ. Health,* 65(1), 38, 1975.
24. **Diaz, L. F., Cooper, R. C., Scarpace, L. P., Trezek, G. J., and Golueke, C. G.,** Public Health Aspects of Composting Combined Refuse and Sludge and of the Leachates Therefrom, final report for the state of California Solid Waste Management Board, June 1977.
25. **Cooper, R. C. and Golueke, C. G.,** Survival of enteric bacteria and viruses in compost and its leachate, *Compost Sci./L and Util.,* 20(2), 26, 1979.
26. **Gorman, P. G., Fiscus, D. E., Schrag, M. P., and Shannon, L. J.,** Comparative Assessment of Bacterial and Viral Field Sampling Methods Used at Solid Waste Handling and Processing Facilities, draft report prepared for the Municipal Environmental Research Laboratory, Office of Research and Development, U.S. EPA, Cincinnati, Ohio, August 1978.

Chapter 8

# LANDFILL—THE ULTIMATE DISPOSAL

## I. INTRODUCTION

Despite the fact that the primary emphasis of this publication is on resource recovery, a section on landfilling is justified, because regardless of the method of processing, a residue remains for which there not only is no further use, but there also is no further applicable treatment other than to return it to the land. Landfilling is the ultimate disposal method in that it is the only one that is applicable to all wastes as well as to waste components from which all utility has been exhausted in a resource recovery operation—in other words, the final rejects. Depending upon the extent of recycling practiced or the methods of processing, theoretically the amount remaining for final disposal may range from 15 to 25% of the original volume. For example, of the destructible fraction, about 15% becomes ash in thermal processing, and as such must be removed and disposed. In practice, however, the ultimate reduction in volume of urban wastes going to the landfill at maximum resource recovery will be about 60%.[1] The lower reduction is the result of two factors: (1) the high compaction ratio of municipal refuse as compared to those of other components of the urban waste stream; and (2) municipal waste constitutes only about 53% of the total urban waste stream. The higher compaction ratio of municipal wastes in relation to that of other urban waste fractions results in a correspondingly lesser reduction in landfill volume requirements brought about by resource recovery. The important conclusion then is that the landfill will always be an essential element in solid waste management.

Landfilling and resource recovery are not necessarily mutually exclusive. Certain resources can be mined at a future time. More practical, however, is the possibility of recovering the combustible gas, methane, which normally is generated within a fill. Landfilling can serve as an interim measure and as such takes care of wastes not amenable to processing by existing systems. Not only is it the most catholic of all disposal systems in its accommodation, it also requires the least outlay in the way of equipment and operation. In view of these characteristics, it is not surprising that landfilling was the first method to be applied to solid waste disposal. If one takes the term in its broadest sense, namely to discard to the land, the practice could be said to date back to man's earliest history.

Relatively speaking, the art of landfill has advanced considerably in recent times, although it still includes much that is primitive. As hinted in the preceding paragraph, the breadth of the meaning of the term "landfill" as used in solid waste disposal extends from the mere dumping of wastes at a specified site to the complete burial and segregation of the wastes and of their breakdown products from the environment.

## II. CLASSIFICATION

Typically, landfills are classified into two types, the *open dump* and the *sanitary landfill*. Excepting for the disposal of certain innocuous materials (e.g., demolition debris), the open dump gradually is being phased out in the U.S. and is being replaced by the sanitary landfill.

### A. Open Dump

Generally the open dump is defined negatively in relation to the sanitary landfill, that is, in terms of what it is not. Thus, according to the terminology adopted by most states and as proposed by the U.S. Environmental Protection Agency, an open dump is a land site which serves as a final depository for solid wastes whether or not the latter are burned or buried, and which does not meet the minimum requirements for a sanitary landfill. The wastes remain exposed to the elements, vectors, and scavengers.

An open dump provides shelter for rodents and serves as their source of food. As a result, an open dump characteristically harbors a large population of rats and is a focal point for their spread beyond the dump. It has been estimated that an area thus affected has about a 5-mi radius. Flies breed and multiply in great numbers in the exposed wastes. Food and beverage containers holding moisture serve as breeding sites for mosquitoes. In addition to the very real hazard to public health coming from the proliferation of rodents and vectors are the two prime affronts to aesthetics, namely unpleasant odors and unsightliness. While the unsightliness may be concealed by fencing or landscaping, the odors cannot be confined. The resulting adverse effect on land values extends well beyond the area immediately adjacent to the dump.

Because of the gravely objectionable consequences of the practice, one would expect that the open dump method had long since been abandoned, or at least be rarely used. The fact is that as late as the mid 1970s, open dumping was the method of solid waste disposal used by at least 80% of U.S. municipalities having a population greater than 5,000. Undoubtedly, the percentage has not been greatly lowered since then. Fortunately, as far as public safety and environmental quality is concerned, the practice is being phased out in response to governmental prodding in the form of regulations, guidelines, and economic incentives.

### B. Sanitary Landfill

#### *1. Definition*

Although many definitions can be found for the term "sanitary landfill", they all have certain elements (key words or phrases) in common. The definition proposed by the U.S. EPA is typical: "A sanitary landfill is a land disposal site at which is employed an engineered method of disposing solid wastes in a manner through which environmental hazards are minimized by spreading the solid wastes in thin layers, compacting the solid wastes to the smallest practical volume, and applying and compacting cover material at the end of each operating day."[2] The key elements in the definition are *compaction, covering with soil each day,* and *minimization of environmental hazards.* The absence of any one of these elements makes a disposal operation anything but a sanitary landfill. Probably the most frequent misuse of the term is its application to operations in which the wastes are not covered each day. Usually the reason for the latter shortcoming is a lack of readily available soil. Regardless of the reason, such sites cannot properly be labeled as sanitary landfills.

#### *2. Types of Sanitary Landfills*

Judging from the literature, the classification of sanitary landfill methods is a rather arbitrary affair. Some authors divide them into only two categories, while others make as many as five divisions. With the former, the distinction is based mainly on whether or not an excavation is made, and hence the two types are the "area" and the "trench" methods. When the division results in five categories, the distinctions are based on excavation or lack of it, on procedures, and on topography. The resulting types are "trench", "area", "ramp", "valley", and "low area". Actually one could extend the divisions until a point is reached at which the exercise not only became unwieldy, but also would be useless. Probably the most convenient division is into "trench" and "area", with perhaps the "ramp" method added to the two, although in reality the latter is a variation of the "trench" and "area" methods.

In the "area" method, the wastes are spread upon the land, are suitably compacted, and are covered with earth. Normally the earth for the cover must be brought in from adjacent areas. The "area" method is best suited for flat or gently sloping lands. The method also can be applied in filling natural and man-made depressions. Examples of the former are ravines, canyons, and valleys, and of the latter, quarries and gravel pits.

As the name implies, with the "trench" method, solid wastes are deposited in a trench specifically cut to receive them. The wastes are layered and compacted, and are covered with earth excavated from the trench. The "trench" method is best applied in flatland areas where the water table is not near the ground surface. An advantage of the method is that the earth excavated from the trench can be used as cover material.

The "ramp" method can be combined either with the "area" or with the "trench" methods. It involves dumping and layering the wastes on the side of an existing slope. The cover material usually is obtained by excavating just ahead of the working face of the daily fill. This method is suited to all areas. Because it can be done with only one piece of equipment, it is particularly adaptable to small operations.

## III. THE SANITARY LANDFILL

### A. Design and Construction

#### *1. Site Preparation*

The principal elements in site preparation are the provision of all-weather access roads to the site, erection of fencing, construction of facilities for personnel and equipment, and installation of weighing facilities and utilities (power and water). The roads should be wide enough to accommodate vehicles as large as transfer trucks, if the use of the latter is anticipated. Fencing serves to conceal the operation from neighbors and passersby, to restrict entry to the site, and to control windblown paper. The concealment can be furthered by suitable landscaping. The type of facilities for personnel and equipment vary according to the size and location of the operation. In all cases, sanitary facilities should be provided. Weighing facilities are necessary for the collection of accurate data on quantities of refuse. These data are needed for cost accounting and good management. The need for utilities is obvious.

#### *2. General Description*

The construction of a sanitary landfill is an incremental affair in that it consists of a series of additions made each day until the full capacity of the site is reached. The overall procedures are illustrated in Figure 1. In the operation, incoming solid wastes are deposited in a series of thin layers superimposed one upon the other. When the first day's input arrives, it is spread in a thin layer and then is compacted by means of a bulldozer or other suitable piece of equipment. As the inflow of wastes continues, a series of layers is built upon the first, each layer being thoroughly compacted, until the total thickness of the layer amounts to 1.8 to 2.4 m (6 to 8 ft). When this depth is reached a layer of earth is spread over the wastes and is thoroughly compacted. This layer of soil is known as the "temporary cover". At the end of the working day, the "face" (exposed working front) of the layered wastes also is covered with soil. The effect of the daily covering of the wastes is the establishment of a cellular structure. In a well-managed operation, a minimum of wastes is exposed at any one time, and the distribution of the wastes is such that the full height of the lift is reached at the end of each day's operation. This procedure is repeated daily until the site is covered. The result is a 1.8- to 2.4-m layer of compacted wastes and cover material that consists of a series of cells as is shown in Figure 1. Such a layer usually is termed a "lift". Where circumstances permit, a second, and in special cases, a third or more lifts may be successively superimposed until

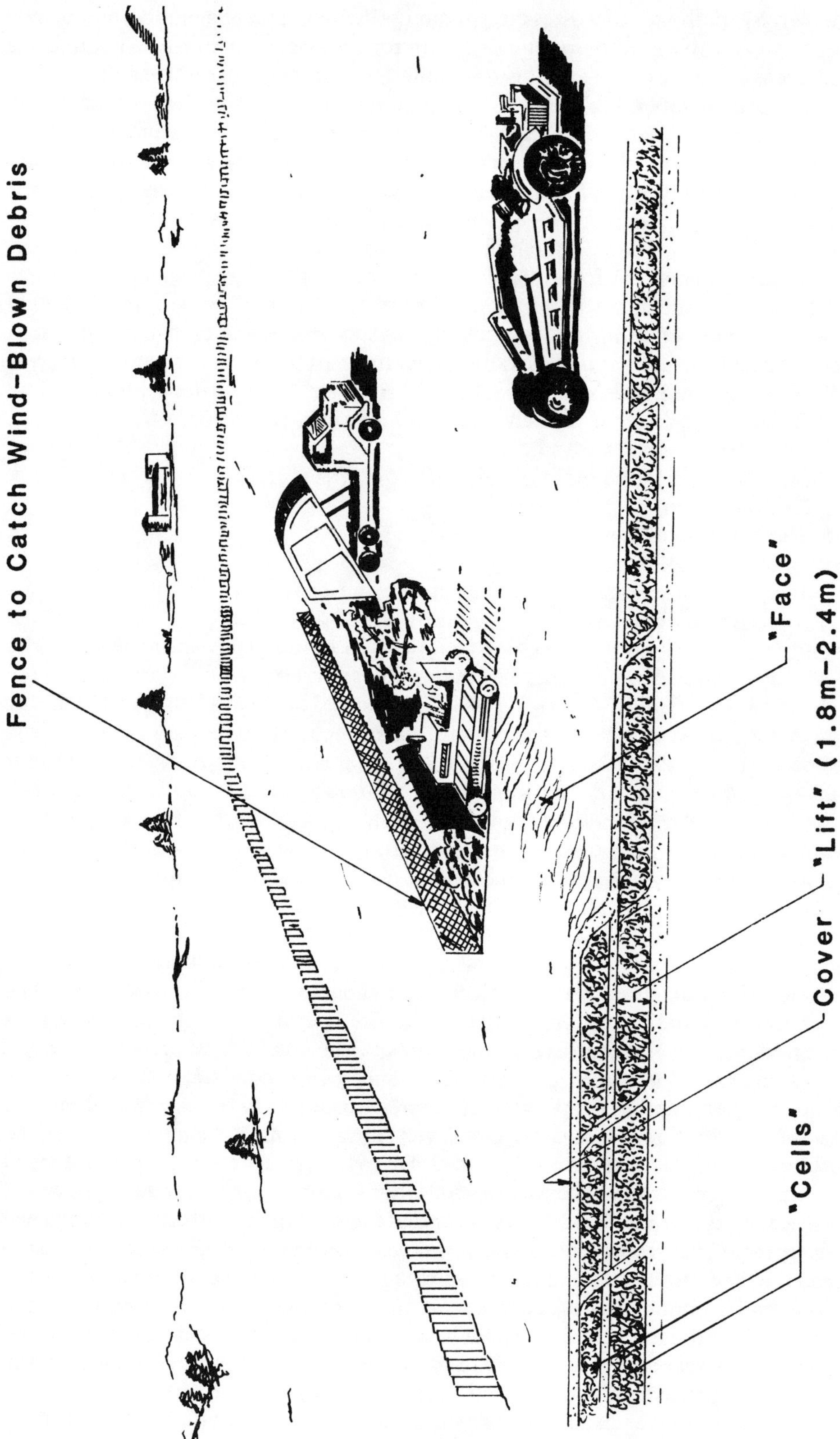


FIGURE 1. Construction of a sanitary landfill.

the full capacity of the site is reached. At that point in time, a final layer of soil is placed upon the temporary soil cover of the topmost lift. The two layers of soil constitute the permanent cover. The site then is designated a "completed landfill".

The number of lifts depends upon the nature of the fill. If the trench method is used, only a single lift may be made, although the addition of a second lift would present no difficulty. On the other hand, if a canyon or ravine is to be filled, the number of lifts is determined by the depth of the depression.

*3. Compaction*

Compacting the refuse is an important step for three reasons: (1) the capacity of the fill is increased in proportion to the degree of compaction; (2) compaction prevents undue settling; and (3) it results in a firm footing for the passage of equipment.

A manifestation of compaction is an increase in the density of the wastes with a corresponding reduction in volume, as is shown by the family of curves in Figure 2. The higher densities graphed in the figure are attained by compacting each layer of incoming refuse as soon as it reaches 15 to 30 cm (12 to 20 in.) with the use of appropriate equipment. Compacting should be done by way of a series of thin layers, because with deep layers, the weight of the compactor is spread over a broader area, and thus the effective pressure per unit volume of wastes is lowered. The usual range of on-site densities after compaction is from 475 to 712 kg/m$^3$ (800 to 1200 lb/yd$^3$).

The reduction in extent of settling is, in turn, important for two reasons. First, excessive settling leads to the development of fissures in the cover, and thus allows exposure of the fill contents to the environment. Secondly, extent and duration of settling have an important bearing on the utility of the completed fill.

With a small operation, compaction is accomplished by a bulldozer. In larger operations, equipment especially designed for compaction is used. An example of such a machine is the steel-wheel compactor commonly called the "sheepsfoot". Other types of compactors are equipped with rubber tires. The "sheepsfoot" has the advantage of incorporating a wearing and tearing action with the concentration of compaction pressure on the cleats. The result is an increase of compaction as much as 40% over that attained with other machines of equal weight. At the time of this writing, the largest such machine constructed for landfilling wastes weighed 35.8 Mg (39.5 tons) and exerted a compaction pressure of 337.2 kN/m$^2$ per wheel tooth (3000 psi).[3] A comparison of various compactors as well as of other equipment is made in Table 1. Regardless of type of equipment, the extent of increase in density brought about by number of passes over a given volume of material diminishes very rapidly after the fourth or fifth pass. Consequently, repetition of passes beyond the fifth becomes a waste of time and money.

Compaction brings with it a potential penalty in terms of rate of stabilization. The more the wastes are compacted, the less the interstitial volume and hence the amount of air trapped in the fill. The partial exclusion of air results in an abbreviation of the aerobic phase of decomposition in the fill. The main disadvantage inherent in the abbreviation is a sharp reduction in the initial rate of decomposition, and hence a prolonging of the time to attain stabilization of the contents of the fill. Aside from the shortening of the aerobic phase, decomposition may be slowed by an inhibition of microbial activity brought about by an excessive concentration of decomposition products and intermediates, because at a constant rate of decomposition, the concentration of product per unit of interstitital volume increases as the volume decreases. In practice the advantages to be gained from compaction far outweigh the disadvantages. Moreover, in operations designed for methane recovery, the anaerobiosis encouraged by compaction becomes a distinct advantage, because oxygen is inhibitory to the methane-forming bacteria.

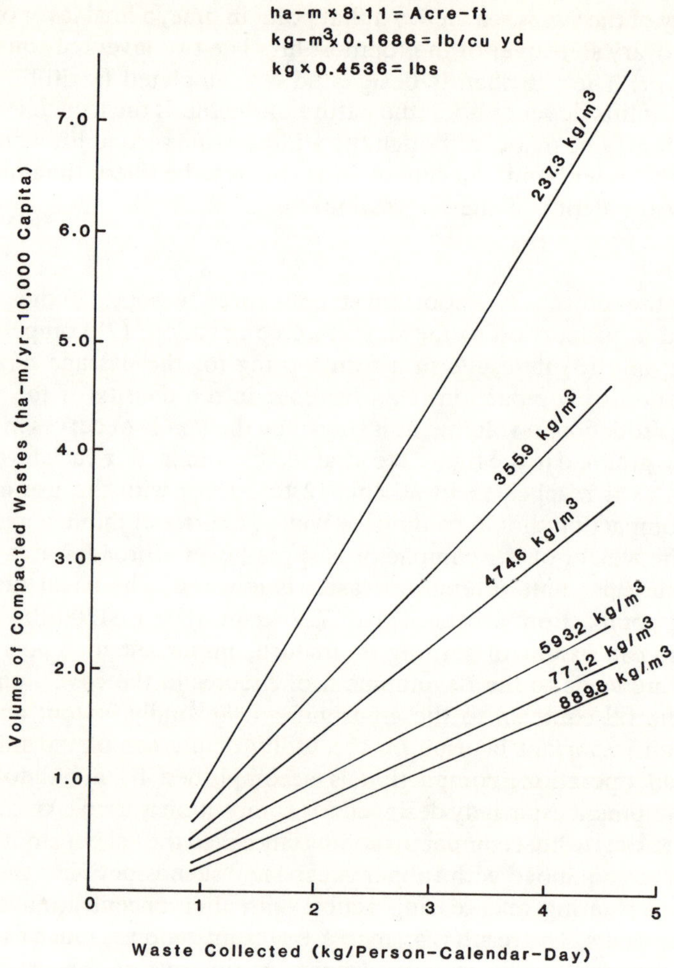


FIGURE 2. Relation between density and volume per hectare per year.

*4. Cover*

The purpose of the cover is to interpose a barrier between the wastes and the environment. The functions of the cover are several: (1) it prevents access of rodents and vectors to the shelter and food offered by the fill contents; (2) it exerts an important influence on the amount of surface water penetration; (3) it prevents methane from escaping the fill; (4) it minimizes the development of visual and olfactory nuisances; (5) it facilitates movement of equipment and vehicles; and (6) it serves as a medium to support plant growth. In summary, it is the presence of a cover, completed daily of course, that distinguishes a sanitary landfill from an open dump.

The type of soil used for the daily cover ideally should be a mixture of clay and sand. Clay alone is unsuitable because when wet, equipment bogs down in it. As it dries, deep fissures are formed that may extend down to the wastes, and thereby breach the integrity of the cover. Sand, on the other hand, cannot be compacted sufficiently to support equipment and methane can diffuse through it. Moreover, larvae and other developmental forms of vectors could readily burrow through such a cover.

**Table 1**
**RATING OF PERFORMANCE OF VARIOUS TYPES OF EQUIPMENT**

| Equipment | Solid waste | | | Cover material | | |
|---|---|---|---|---|---|---|
| | Spreading | Compacting | Excavating | Spreading | Compacting | Hauling |
| Crawler dozer | E | G | E | E | G | NA |
| Crawler loader | G | G | E | G | G | NA |
| Rubber-tired dozer | E | G | F | G | G | NA |
| Rubber-tired loader | G | G | F | G | G | NA |
| Landfill compactor | E | E | P | G | E | NA |
| Scraper | NA | NA | G | E | NA | E |
| Dragline | NA | NA | E | F | NA | NA |

*Note:* Basis of evaluation is easily workable soil and cover material, and haul distance greater than 300 m (1000 ft).

Rating key: E = excellent;
G = good;
F = fair;
P = poor;
NA = not applicable.

Modified from Brunner, D. R. and Keller, D. J., Sanitary Landfill Design and Operation, (SW-65ts) U.S. EPA, 1972.

Consequently, a proper mixture is one that can be thoroughly compacted, that supports the passage of equipment under all weather conditions, and in which fissures do not develop. The type of soil to be used for the final or "permanent" cover depends upon the use to be made of the completed fill. For example, if the fill is to be used as a "green space" the soil should be one that can support plant growth.

The depth of the cover depends upon its position and the phase of the operation. During the time the fill is active, the thickness of the top cover placed each day ("temporary cover") should be about 30 cm (12 in.) after compaction. The depth of the layer on the face (working face) at the end of the day should be from 15 to 20 cm (6 to 8 in.). The respective thicknesses are those proven by experience to be suitable to accomplish the purpose of the fill.

The minimum depth of the final cover should be 30 cm (12 in.). Therefore the combined (total) depth of the permanent cover is 60 cm ( 24 in.). For special applications the final cover may have to be deeper. For example, if tall shrubbery and trees are to be planted, the depth of the permanent cover should be great enough to accommodate a root growth that does not come in contact with refuse or with gaseous decomposition products. Carbon dioxide and methane may inhibit root hair development through the exclusion of atmospheric oxygen, and also through the formation of carbonic acid in the presence of water. The potential presence of substances toxic to plants is another factor.

Probably the most complete and authoritative report or manual on landfill covers available now is the publication by R. J. Lutton et al.[5]

## IV. PROCESSES THAT TAKE PLACE IN A SANITARY LANDFILL

The processes or reactions which take place in a sanitary landfill have an extremely important bearing upon the environmental impact of the fill as well as upon its utilization when completed. The processes are of three types: physical, chemical, and biological. Of the three, the biological processes are the most significant.

**A. Physical**

Physical reactions affect all components of the buried wastes. The reactions are in three broad forms, namely compression and settling, dissolution and transport, and sorption. In addition to the compaction imparted by machinery during the active stage of the fill operation, further compression takes place after the wastes are in place due to the weight of the wastes and of the soil cover. Moreover, a certain amount of consolidation results from the sifting of soil and other fines. The end result is a further settling of completed fill exclusive of that resulting from other reactions.

The extent of the second set of physical processes depends not only on that of chemical and biological reactions but also upon the amount of water that may enter the fill. The water forms a medium for the solution of substances subject to dissolution and for the transport of unreacted materials. The latter may be in the form of living and nonliving particulates ranging in size from colloidal to several millimeters in cross section.

Although at present, methods are not available for fully determining the extent of adsorption in the fill itself, it is certain that because of the broad variety of components and particle sizes of the wastes, a considerable amount does take place. The environmental significance rests upon the fact that adsorbed material is immobilized within the fill and therefore does not enter the environment. For example, adsorption should play an important part in the containment of viruses. A problem is the questionable permanency of the adsorption resulting from, among other factors, the effect of chemical and biological decomposition on the number of adsorption sites.

The major potential for absorption as far as municipal refuse is concerned exists in the form of the paper and paper products, that is, the paper fibers of which these products are constituted. Although there are other fibers (cotton and wool) in refuse, their amounts border on the insignificant. The significance of absorption is not only the immobilization of potential pollutants within the fill, but more importantly the immobilization of the water that serves to transport pollutants out of the fill. Eventually, however, all absorbent material in the fill becomes saturated, and so in some respects absorption may be regarded only as a delaying action.

**B. Chemical**

Chemical reactions are mainly in the form of oxidation and of those made possible by the organic acids and carbon dioxide synthesized in the biological processes. The effect of the carbon dioxide depends upon its being dissolved in water. The oxidation reactions are rather limited in scope and are dependent upon the supply of oxygen trapped at the time the fill was being made. Components most likely to be affected are the ferrous metals.

Reactions made possible by the synthesis of organic acids and the dissolution of carbon dioxide in water are typical acid-metal reactions. The manifestation of these reactions are the metallic ions and salts found in the liquid contents of the fill. The environmental significance results from the fact that nonpolluting materials are solubilized and thus are transformed into potential contaminants. When carbon dioxide dissolves in water, the quality of the latter is thereby deteriorated, especially where the substrata contain calcium and magnesium.

**C. Biological**

In terms of treatment, the importance of the biological processes derives from the fact that through them the organic fraction is rendered biologically stable. It is converted into a state in which it does not constitute a potential source of nuisances. A more tangible result is the reduction in volume of the organic fraction, inasmuch as a sizeable portion of the carbonaceous components is converted into gas. The collection of biodegradable material includes not only the garbage fraction, but also paper and paper products and other fibrous material of plant or animal origin. Inasmuch as the biodegradable portion constitutes the major share of the buried wastes, it becomes obvious that the impact of biological decomposition is an exceedingly important one. Although the organic fraction serves as the primary substrate for the biological reactions, some inorganic components also may be affected indirectly.

*1. Aerobic Decomposition*

Directly after the wastes are buried, decomposition takes place primarily under aerobic conditions and continues to do so until the oxygen content of the interstitial air is depleted. The duration of the aerobic period is rather brief and is determined by the degree of compaction of the material and by the moisture content. Organisms that are involved range from obligate aerobes to facultative anaerobes. Types of microbes include bacteria and fungi and to a limited extent, protozoa. Inasmuch as the decomposition is primarily aerobic, the temperature of the fill contents rises and may reach levels approaching 45 to 55°C. The adverse environmental impact of the aerobic decomposition is relatively minimal because the ultimate end products of biological aerobic breakdown are ash, carbon dioxide, and water. In reality, intermediate breakdown products undoubtedly are released, but their extent and pollution contribution are slight. Another point to remember is that a certain fraction of the nutrient elements are transformed into microbial protoplasm, and thereby make up a reservoir of organic material for future decomposition.

*2. Anaerobic Decomposition*

With the dwindling of the oxygen supply comes a gradual shift to anaerobic decomposition. Indeed, by far the greater part of the decomposition of the organic matter occurs under anaerobic conditions. The activities taking place in this stage are ones which lead to conditions that have a potentially highly unfavorable impact on the environment.

The anaerobic decomposition is of the same nature as that in the anaerobic digestion of sewage sludge. The principal difference is that the latter is controlled to encourage methane production, whereas no control is exercised over the reactions in a fill. Types of organisms include facultative and obligate anaerobes. Among the latter are the methane-producing bacteria. Since almost all fungi are aerobic organisms, the fungal population is negligible. Unlike the aerobic phase, no discernible rise in temperature takes place. Consequently, as the shift from aerobic to anaerobic decomposition proceeds, the high temperature characteristic of the aerobic phase drops to an approximation of that of the ambient.

The decomposition products of interest as far as environmental impact is concerned fall into two main classes, namely the volatile acids and the so-called biogases. The volatile acids produced in digestion are short-chain fatty acids, and include formic, acetic, propionic, butyric, valeric, caproic, and isovaleric acids. An important characteristic of this assortment of acids is an extremely objectionable odor. The acids serve as a substrate for the methane-producing microorganisms.

The two principal gases formed in anaerobic decomposition, and hence in the fill, are methane and carbon dioxide. In addition, hydrogen sulfide, hydrogen, and nitrogen may be found in trace amounts, that is at a combined total of 5% or less. The ratio of methane to carbon dioxide not only is a function of the size and activity of the methane formers in relation to those of other microbial forms (primarily acid formers), but also of the amounts and nature of the carbonaceous components of the wastes. For example, if the organic fraction were entirely cellulosic, the ratio of methane to carbon dioxide would be approximately one. The addition of proteins and fats tips the balance in favor of methane. With the proportion of protein and fats to cellulose as found in municipal refuse, the ratio of methane to carbon dioxide should be about 1.1. to 1.2. The volume of gas to be expected depends upon the extent and rate of decomposition. Experience with the digestion of municipal refuse in a reactor under operating conditions conventionally provided in digesting sewage sludge inicates that a yield of about 0.25 to 0.37 $m^3$ (6 to 7 $ft^3$) of gas could be expected for each kilogram of volatile matter contained in a fill. However, because of the less than favorable conditions prevailing in a fill, such yields would be the upper level of production. Experience thus far indicates a range of about 30 to 94 $m^3$ methane/Mg refuse (1000 to 3000 $ft^3$/ton).

*3. Environmental Factors*

Factors which affect the rate and extent of biological decomposition in a fill are those which apply to all biological activities. Key factors as far as the average fill is concerned are moisture, temperature, and the degree of resistance of the substrate (wastes) to bacterial attack.

If the carbon to nitrogen ratio (C/N) of the refuse is above 45 to 50:1, the proliferation of the acid-forming bacteria takes place at the expense of the methane formers. The result is a potential accumulation of volatile acids great enough to inhibit all biological decomposition. The effect of resistance is obvious. If the wastes are in forms resistant to bacterial attack (e.g., of lignin), the nutrients locked in the resistant material remain unavailable to the microorganisms. A third factor respecting nutrient has in recent years received attention, namely the effect of size reducing the wastes on extent and rate of decomposition. Undoubtedly, other conditions remaining unchanged, microbial activity would be enhanced by reducing the particle size of the wastes.

Moisture is essential to bacterial survival and hence to rate and extent of decomposition. An ideal moisture content is one that approaches saturation. However, under conditions prevailing in a landfill, moisture content probably would not become limiting until it dropped below about 55 to 60%. All biological activity ceases when the moisture drops to about 12%. Certainly, the rate of decomposition has slowed to extremely low rates long before the 12% level is reached. Consequently, in the U.S. decomposition proceeds extremely slowly in the southwestern tier of the states.

Temperature ranks with moisture content in terms of importance in that rate of biological activity increases with rise in temperature, at least within the mesophilic range of 5 to 40 or 45°C. Therefore, one would expect decomposition to proceed rapidly in the southern states. Inasmuch as the favorable temperature is counteracted by low moisture content in the southwestern states, in terms of conditions favorable to decomposition, the southeastern states are the most favcrably situated in the U.S.

## V. ENVIRONMENTAL IMPACT

### A. Leachate Production and Control

Perhaps the greatest potential environmental impact exerted on the aqueous environment by the various processes in a sanitary land fill is that of leachate formation. Leachate production is a means of externalizing certain of the processes described in the preceding paragraphs.

The term "leachate", as used with respect to sanitary landfills, is based upon the dictionary definition of the term but adapted to fit the landfill situation. Thus, it may be defined as the water and its burden of dissolved and suspended solids resulting from the percolation of water through a landfill, sanitary or otherwise. Inasmuch as the characteristics of a leachate are those of the substances making up the leachate, its

**Table 2**
**COMPOSITION OF LEACHATE**

| Item | Leachate | | Raw sewage[a] | |
|---|---|---|---|---|
| | Year 1 (mg/ℓ) | Year 2 (mg/ℓ) | Strong (mg/ℓ) | Weak (mg/ℓ) |
| Total solids | 45,070 | 13,629 | 1,793 | 796 |
| Suspended solids | 172 | 220 | 1,190 | 640 |
| Dissolved solids | 44,900 | 13,409 | 603 | 156 |
| Total hardness (as $CaCO_3$) | 22,800 | 8,930 | 339 | 204 |
| Calcium (as $CaCO_3$) | 7,200 | 216 | 239 | 137 |
| Magnesium (as $CaCO_3$) | 15,600 | 8,714 | 100 | 67 |
| Total alkalinity ($CaCO_3$) | 9,680 | 8,677 | 335 | 245 |
| Ammonia (N) | 0.0 | 270 | 53.3 | 23.4 |
| Organic N | 104 | 92.4 | 24.6 | 19.2 |
| BOD (O) | 10,900 | 908 | 538 | 385 |
| COD (O) | 76,800 | 3,042 | 957 | 329 |
| Sulfate ($SO_4$) | 1,190 | 19 | 225 | 81 |
| Total phosphate ($PO_4$) | 0.24 | 0.65 | 11.7 | 7.2 |
| Chloride (Cl) | 660 | 2,355 | 312 | 97 |
| Sodium (Na) | 767 | 1,160 | 267 | 100 |
| Potassium (K) | 68 | 440 | 24 | 12.3 |
| Boron (B) | 1.49 | 3.76 | 0.54 | 0.43 |
| Iron (Fe) | 2,820 | 4.75 | 0.66 | 1.12 |
| pH | 5.75 | 7.40 | 8.05 | 7.40 |

[a] Los Angeles County Sanitation Districts—1971.

nature and strength varies with the composition of the wastes in the fill and with the age of the fill. The solids content of leachate includes dissolved and suspended inorganic and organic solids, colloids, and microbial organisms. The data in Table 2 are representative of the composition of leachate and of the effect of time, and also serve for a comparison between the respective strengths of leachate and raw sewage. According to the data, the early leachates from the landfill far exceeded even the strong sewage in pollution potential, and continued to do so after the expiration of 3 years.

Heavy metals are attracting a considerable amount of attention because of their pollution potential. Since normally the heavy metal content of municipal solid waste is not large enough to be cause for concern, that of the leachate should not be excessive. However, if industrial wastes containing heavy metals are landfilled along with municipal solid wastes, the situation changes as far as leachates are concerned. In a study reported by him, Streng[6] made the following observations:

1. The moisture adsorption capacity of the municipal waste mass was reduced by the addition of semisolid industrial wastes. As a result, the concentration of contaminants in the leachate was materially increased.
2. Metallic ions may become very soluble because of the existence of acid/reducing conditions.
3. The ions may initially be discharged because of cations already solubilized. If not already solubilized the ions are easily leached during initial placement ICu.
4. Ions, such as Hg, may be either chemically or microbially converted into a form that is amenable to transport.

The extent of the time period during which leachate constitutes a potential source of water pollution depends upon the rate of decomposition of the wastes in the fill, and

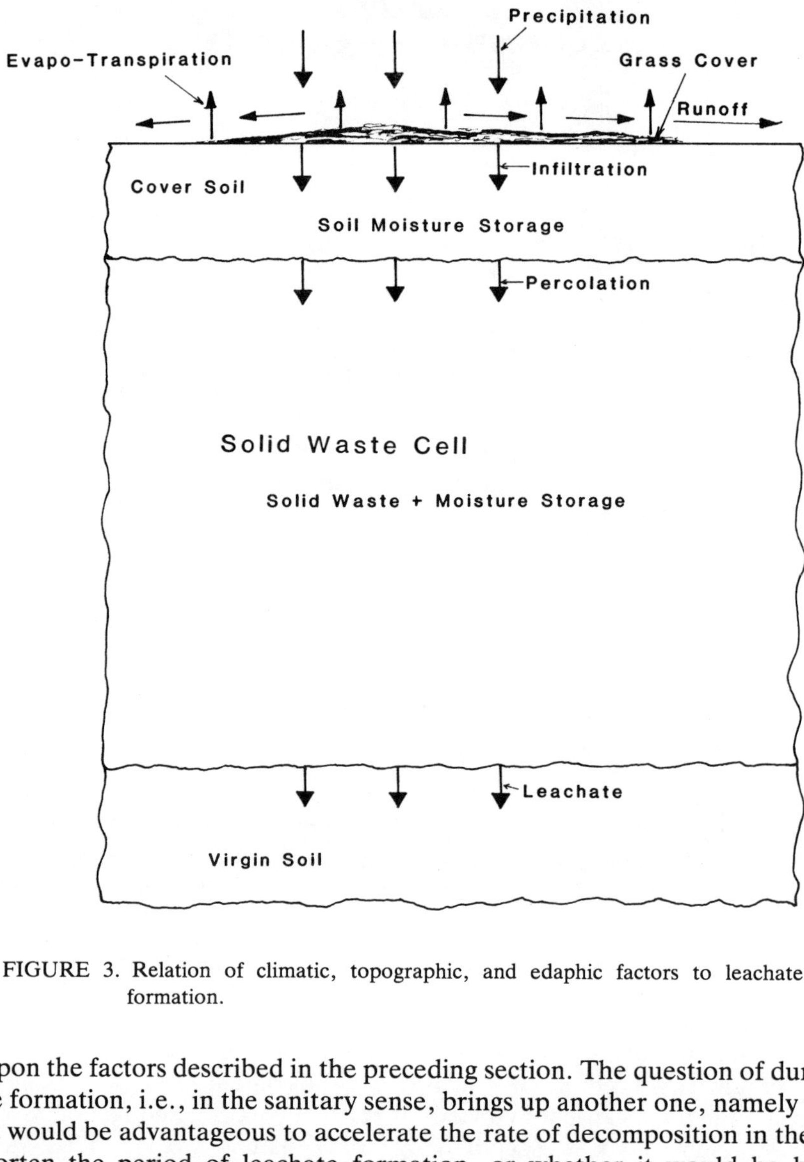


FIGURE 3. Relation of climatic, topographic, and edaphic factors to leachate formation.

hence upon the factors described in the preceding section. The question of duration of leachate formation, i.e., in the sanitary sense, brings up another one, namely whether or not it would be advantageous to accelerate the rate of decomposition in the fill and thus shorten the period of leachate formation, or whether it would be better to concentrate on minimizing the amount generated. Unfortunately, the question remains to be resolved.

The source of the water in leachate is the rainfall that penetrates the fill and the capillary water of the soil. With improperly designed and sited fills, ground and surface waters are additional sources. The water is the runoff from the surrounding terrain. The amount of leachate formed is a function of volume of water that reaches and passes through the fill. Therefore, the total volume of leachate leaving a fill is equal to that of the total precipitation (rainfall) plus that of water from other sources minus the combined volume of the water lost by runoff, evaporation, and plant transpiration, as well as that retained in the fill by absorption. The last named factor ceases to be of significance after the fill has reached field capacity, that is, after the fill has become saturated. The relationship between the various factors is illustrated in Figure 3. The amount of leachate generation to be expected can be determined through a water balance method for prediction described in a report by the U.S. EPA.[7]

**Table 3**
**EFFECT OF SOIL AND CONTOUR ON RUNOFF**

| Surface | Slope (%) | Type of soil (planted with grass) | Runoff coefficient |
|---|---|---|---|
| Flat | 2 | Sandy | 0.05—0.10 |
| Average | 2—7 | Sandy | 0.10—0.15 |
| Steep | 7 | Sandy | 0.15—0.20 |
| Flat | 2 | Heavy | 0.13—0.17 |
| Average | 2—7 | Heavy | 0.18—0.22 |
| Steep | 7 | Heavy | 0.25—0.35 |

**Table 4**
**EFFECT OF COVER SOIL TYPE AND AMOUNT OF WATER THAT REACHES THE FILL CONTENTS**

| Type of soil | Quantity of water ($m^3$/ha per season) |
|---|---|
| Uniform coarse sand | $4.06 \times 10^6$ |
| Uniform fine sand | $4.06 \times 10^4$ |
| Well-graded silty sand and gravel | $4.06 \times 10^3$ |
| Silty sand | 914.4 |
| Sandy clay | 48.62 |
| Clay | 0.91 |

Modified from Salvato, J. A., Wilke, W. G., and Mead, B. E., *J. Water Pollut. Control Fed.*, 43(10), 2084, 1971.

Inasmuch as nothing can be done about rainfall and very little about evaporation, the only practical means of lessening the amount of leachate is to increase runoff from the fill site and to promote transpiration. Runoff from surrounding terrain into the fill can be diverted by interposing intercepting ditches at appropriate locations. Ground water may be excluded through the placement of intercepting drains. Runoff from the fill can be facilitated by covering the waste with a relatively impermeable soil suitably contoured. The influence of impermeability of cover material is illustrated by the data in Tables 3 and 4.

Transpiration refers to the loss of water vapor from living plants. The loss may take place from any green part of the plant exposed to air. By far the greater part of the loss is through pores (stomata) in the leaves of the plant. The usual rate of transpiration is from 5 to 25 g water per square meter of leaf surface. Inasmuch as the water has been removed from the soil by the plant, the plant serves as a conduit for water from the soil to the atmosphere. The total amounts can be substantial, e.g., 1542 $m^3$ (1.25 acre-ft) per growing season with maize, 493 $m^3$ (0.4 acre-ft) with elms, and 2899 $m^3$ (2.35 acre-ft) with red maples. This rate of loss exceeds that from bare soil because the plant acts as a pumping mechanism, and the size and arrangement of the stomata increase the evaporation efficiency of the leaves. Consequently an excellent way of lessening the entry of water into the fill is to cultivate some type of crop on the cover.

In the drier regions of the country, a well-designed combination of transpiration and runoff may lower the volume of water entering the fill to a level at which absorption ties up surplus water without reaching the point of saturation.

Reducing the amount of leachate generated and accelerating the rate of decomposition are almost mutually exclusive accomplishments. To minimize leachate production necessitates a reduction in the amount of water entering the fill, whereas a maximum rate of decomposition involves the provision of an abundance of water. An approach to accomplishing the two aims simultaneously is to collect the leachate and recirculate it through the fill. While strictly speaking, leachate production is enhanced initially by so doing, it is not discharged. Instead, upon being recirculated, it provides the moisture needed for the later stages of decomposition. Moreover according to the proponents of the practice,[9,10] the recirculated leachate theoretically leads to a more rapid development of an active population of methane formers, as well as serves to moderate unfavorable conditions that may exist in the fill. A very distinct advantage of the system is that it serves as a means of treating leachate by repeatedly exposing it to bacterial action within the fill. As of this writing, the long-term practical feasibility of the system remains to be demonstrated. Difficulties involved are mainly in the handling of the leachate and in the reintroduction of it into the fill.

### *1. Fate of Leachate After Leaving the Fill*

If there is an adequate distance between the bottom of the fill and the ground water table, the leachate usually is attenuated considerably before it reaches the ground water table or escapes into a body of surface water through lateral movement. The effect of attenuation is the placing of an upper limit upon the distance a given contaminant travels through the soil.

Attenuation takes place through five major mechanisms: (1) mechanical filtration; (2) precipitation and coprecipitation; (3) sorption; (4) dilution and dispersion; and (5) microbial activity. It takes place during both vertical and lateral movement.

Mechanical filtration is the physical restriction placed by soil upon the flow of suspended contaminants. With respect to leachate, the soil layer between the bottom of the fill and the ground water table and, to some extent, the surrounding soil accomplish a certain amount of filtration. The effectiveness of the soil layer in straining or removing particles from suspension is dependent upon the type of soil (e.g., texture) and the thickness of the layer. The finer the texture, the more effective is the straining action. Thus a clay soil is the most effective of the soils, whereas a loose sandy-gravelly layer would be the least effective. Depth is a relative factor in that more suspended particles are removed within a given distance of movement through a clay column than through a sand column. Filtration is especially effective in limiting the travel of microorganisms. Hence, if the layer of soil is sufficiently thick, the removal of pathogens can be complete. However, the effectiveness in the removal of viruses remains to be satisfactorily demonstrated.

Through precipitation and coprecipitation brought about by changes in temperature, pH, solution composition, and a variety of other conditions, as the leachate moves through the soil soluble compounds may be immobilized by being converted into insoluble compounds. These mechanisms are especially effective in the removal of heavy metals. Under sorption are included the processes of adsorption, absorption, and ion exchange. In addition to the soil particles, organic compounds in the soil, microbial growths, and chemical precipitants all serve as sorption media. The reactions involved in attentuation by sorption are very complex and await further elucidation. Factors demonstrated to be effective are size of soil particles (or total clay content), free hydrous oxides, soil pH level, and solution of flux through the soil.[11] Of lesser importance are sand (lacks exchange capacity), cation exchange capacity, and soil organic matter.[11] The exception with respect to the latter is the immobilization of mercury. Organic matter may also effect a certain amount of chelation. The effect of dilution and dispersion is a decrease in the concentration of the contaminants by the microbial population indigenous to the soil. A ranking by relative degree of attenuation order of elements as leachate passes through clay soils as proposed by Griffin and Shimp[12] is as follows:

| | |
|---|---|
| High | Hg, Pb, Zn, Cd |
| Moderate | Si, Mg, K, $NH_4$ |
| Low | Na, COD, Cl |
| No attenuation | Ca, Fe, Mn |

In their analyses of leachate that had passed through sand columns, Griffin and Shimp found no detectable Pb or Hg and markedly reduced concentrations of Zn and Cd. Apparently the important attenuation factor in the columns was the precipitation of heavy metals hydroxides and carbonates.

Attenuation through dilution and dispersion refers to the reduction in concentration of contaminants resulting from intermixing with the soil water. The degree of attentuation thus accomplished can be significant in the soil at less than field capacity. Farquhar and Rovers[13] found that the dilution rates in three types of soil studied by them ranged from 27.89 mg/g (dry soil) to 110.84 mg/1000 g (dry soil). The dilution rate was highest in the finer soils and least in the coarsest soil.

Attenuation through microbial activity is accomplished through the uptake and utilization of inorganic and organic contaminants by the soil microbial community. Microbial activity may exert an effect indirectly through the synthesis of intermediates and production of end products (e.g., organic acids) that may affect the pH level of the liquid medium. The activity of the indigenous microbial populations contributes to the destruction of pathogenic microorganisms that may be present in leachate. The destruction is the result of a combination of competition, breakdown of nutrient sources essential to pathogens, and antibiosis.

*2. Control of Leachate Travel*

Two approaches can be followed to impede the movement of leachate beyond the immediate confines of the fill: (1) select a site at which topographic, hydrologic, and geologic factors are such as to effectively prevent an appreciable movement of leachate; and (2) Provide an engineered barrier to the movement. Although the first approach is the ideal one, it often cannot be implemented, in which case the second approach must be followed.

*3. Site Selection with Reference to Leachate*

Factors relating to leachate management constitute only a part of the total complex of factors to be considered when selecting a site for a landfill. In fact, any one or more of the nonleachate factors may and usually does override those pertaining to leachate containment. Nonleachate factors are those which make a given location more suitable with respect to collection and transport of wastes, to land use, and to socio-political considerations.

To select a site that ensures maximum natural containment of leachate, the interrelation between three types of factors must be considered, namely, topograpical, hydrological, and geological. The interrelation is well illustrated in Figure 4. Judging from the figure, an ideal site is one on fairly level ground, at which the top soil layer consists of clay extending well beyond the projected depth of the fill, such that a sizeable layer of clay exists between the bottom of the fill and the top of the water table.

The extent of the measures to be taken in the absence of an ideal site depends upon

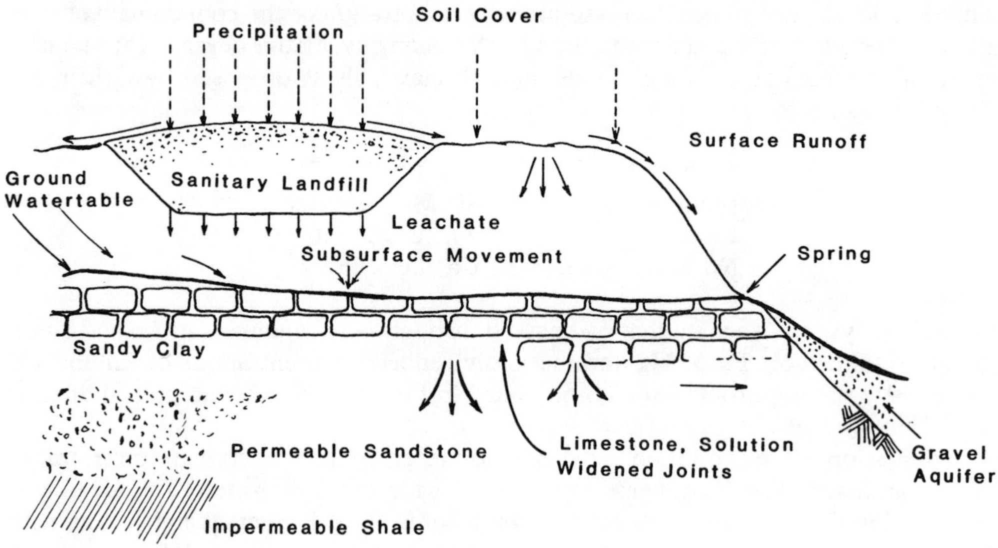


FIGURE 4. Interrelation between topographic, hydrologic, and geologic factors in terms of leachate travel.

that of the departure of the site from the ideal. It is important that any measures that must be taken be incorporated into the design of the fill, because the recourses for dealing with leachate control after a fill is constructed are expensive and difficult to apply. They are to sink wells at strategic locations and pump out the leachate, and to excavate the buried wastes and take the requisite measures or move the wastes to a better site.

With a new construction, the first step is to line the area destined to receive the wastes with an impervious material. While a variety of potential liner materials is available, much remains to be learned about short- and long-term durability. Potential liner materials are clay, bentonite, soil cement, asphalt, and an assortment of membrane liners such as PVC, polyethylene, and chlorinated polyethylene. While some materials have been tried with landfills, the time of construction of the fills is too recent to permit firm conclusions as to the effects of prolonged exposure. Conditions peculiar to a landfill can exert stresses on the liner materials not encountered in other applications. In addition to a mechanical rupture due to excessive unsupported weight and puncture by sharp objects, there is the possibility of a partial or complete disintegration of the liner material because of chemical reaction with the leachate itself. Rupture can be prevented by placing a supporting layer of soil under and on top of the liner. Chemical disruptive action is more difficult if not impossible to counteract.

When leachate is not permitted to leave the fill or the immediate vicinity thereof, provision must be made to collect and remove it for treatment. A suggested arrangement of liner and collection pipes is shown in Figure 5.

The problem of treating leachate awaits satisfactory solution. For the present, where it is feasible, discharge into a sewer for processing at a sewage treatment plant is the preferred course. Normally, the chances of a leachate having an adverse effect on the performance of the plant are slight. The main limitation would be a matter of strength of leachate vs. that of raw sewage. A general rule of thumb seems to be that no problem will be encountered if loading in the form of leacate does not exceed 5% of the total loading to the plant. With one exception, methods that have been proposed for leachate treatment were designed originally for conventional sewage treatment. Of the latter,

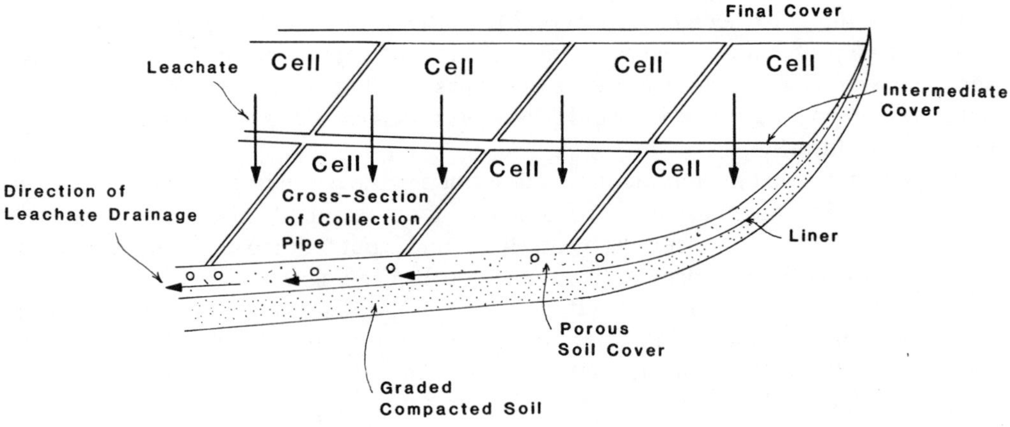


FIGURE 5. Arrangement of liner, drains, and collection pipes in a sanitary landfill.

lagooning and spray irrigation offer some promise. However, all of the presently conventional methods are attended by one or another difficulty when applied to leachate treament. The exception is the previously mentioned recycling of leachate within the fill.

**B. Gas Management**

Because economic reality (i.e., practicality) permits very little to be done in the way of controlling the rate and amount of gas production, the primary effort in the management of gases formed in a fill is to control their movement. The amount of effort to be thus expended is a function of the potential environmental impact of the gases. An exception is constituted by those instances in which provisions are made for utilizing the methane fraction as an energy source. Considerations regarding the environmental impact depend upon the use to be made of the completed fill and on the proximity of the fill to residential areas.

Of the gases produced, the two of major concern are methane ($CH_4$) and carbon dioxide ($CO_2$). The impact of other gases that are generated is the production of an objectionable odor where the amount escaping into the environment is significantly large. Generally the concern about the carbon dioxide fraction is less than that about the methane. The main effect of carbon dioxide is exerted when it dissolves in water to form carbonic acid. As such it can dissolve minerals and thus be instrumental in promoting the travel of the latter to the ground water, and if the subsoil contains calcium and magnesium, the hardness of the water is increased. At high concentrations in the root zone, carbon dioxide could displace oxygen or bring about an unfavorable shift in pH and thereby interfere with plant growth.

If the methane produced in a fill were completely unconfined, that is, immediately dissipated into the atmosphere as rapidly as it were generated, then the gas would pose no environmental problem. The reasons are threefold: (1) it is nontoxic; (2) it is odorless, tasteless, and colorless; and (3) it is generated at a rate such that the concentration at any instant is not great enough to pose a hazard even if it were toxic. The situation changes when conditions are such that the gas accumulates within a confined area, as is the case in a landfill. Then the gas becomes a hazardous substance, primarily because it is combustible, and when mixed with air at a concentration of 5 to 15%, it becomes explosive. At high concentrations it also becomes an asphyxiating agent in that it displaces a certain amount of oxygen ($O_2$). Even with the two potentials (explosivity and displacement of oxygen), methane production and travel would constitute no problem if no use were intended for the completed fill and no dwellings were nearby. It is true that the presence of the gas might limit root penetration by cover plants, but in time this limitation would disappear because of the slow dissipation of the gas through diffusion.

The problem is that the ideal situation described in the preceding paragraph rarely is encountered. Because of urban sprawl and need for maximum land use, the completed fill eventually becomes surrounded by dwellings and is put to some use. Consequently it is the part of wisdom to provide for the control of methane.

The collective phenomena and systems involved in the movement of gases through soil are extremely complex in nature. For practical purposes it can be said that they migrate laterally and vertically from the fill, and follow the path of least resistance in so doing. Thus they will travel more rapidly and further through loosely textured soil than through a heavy compacted soil. In installations surrounded by permeable soils, the distance traveled in later migration depends upon the permeability of the ground surface. The maximum radial extent of methane at a 5% concentration above the elevation of the base of the fill extends from 1.28 to 2.25 times the radius of the fill.[14] Actual distances of 180 m (600 ft) from a fill have been reported.[15] If the fill were completely surrounded by a saturated compact layer of clay, methane would not escape. Compacted clay physically impedes the passage of any gas. With methane, an additional factor comes into play, namely, it is practically insoluble in water. Therefore, a twofold barrier exists in the form of water and clay. The saturation would change the situation somewhat with respect to carbon dioxide, however. Carbon dioxide being soluble in water could diffuse through the clay barrier, but then it would no longer be a gas.

An obvious method of preventing either an accumulation of gas in the fill or lateral travel from the fill is to vent it to the atmosphere. Vents are of two kinds: layers of gravel appropriately placed and wells sunk into the fill. Gravel vents may be constructed in the form of a 30.5- to 45-cm (12- to 18-in.) gravel layer placed upon the cover of the face of individual cells in a landfill as in Figure 6A. A second form of gravel vent is a circumferential trench excavated to the depth of the fill and filled with gravel (Figure 6B). Gas-venting wells may be equipped with waste gas burners or for gas collection and storage. The efficiency of gas wells can be increased by pumping out the gases at a slightly negative pressure.

Despite the presence of wells and gravel vents, some lateral movement does take place. Moore[14] found that while vents may be useful in reducing methane concentration along a radial line passing through the vent, they are essentially useless in reducing concentrations along radial lines a few arcs away. In the absence of an impermeable barrier aroud the fill, probably the most effective way of preventing or at least sharply limiting lateral movement is through the use of the circumferential trench. The deeper the trench, the more effective it becomes.

A solution to the problem of methane control could be to eliminate the problem at its source by preventing methane generation. This can be done by introducing reagents or substances inhibitory to the growth of methane-forming bacteria into the fill along with the waste. Such an approach is not without serious difficulties. For example, there is the cost of the reagents. If the reagent is a part of a waste not normally landfilled along with refuse, then there is the cost of transporting the toxic waste to the landfill site plus the resulting accelerated filling of the available landfill capacity. A complication might come from the reagent or waste becoming a water pollutant. Then, there is the matter of insuring a thorough mix between refuse and reagent. Unless each and every particle of waste contacts the reagent, some methane generation will take place, and hence the

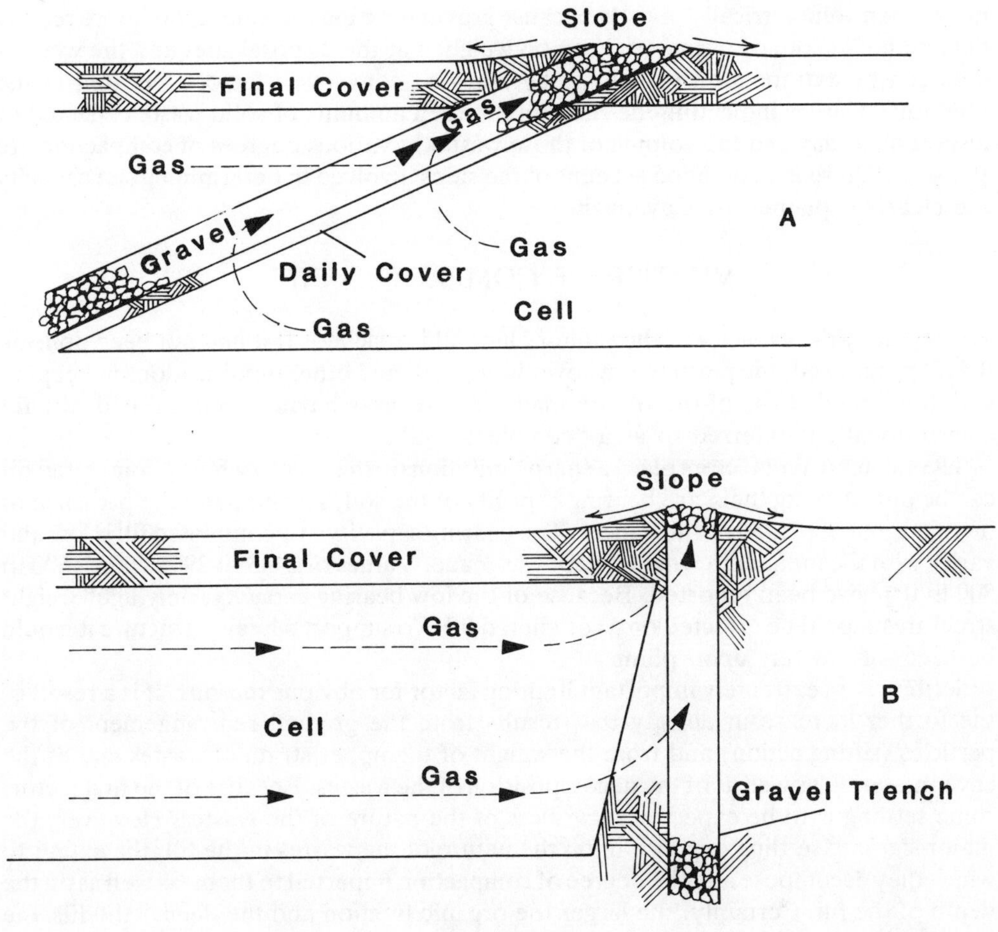


FIGURE 6. Dissipation of gas from a landfill.

need for providing another control continues to exist. Accomplishing such a mix would not only be expensive, but it would also disrupt the landfill operation. Yet another disadvantage is that the output of leachate could be increased. The reason lies in the fact that the organic acids formed by acid-producing organisms serves as the substrate for the methanogens. If the methanogens are killed off, obviously the organic acids will remain. Since it is much easier to treat methane than to treat leachates, the trade-off hardly seems worthwhile.

## VI. SITE CAPACITY—LAND REQUIREMENTS

The expected life span of a sanitary landfill is a function of a collection of factors that becomes obvious with a minimum of reflection. Any one of the factors can be a basic factor. The first factor is the volume of the fill. The others are rate of generation of wastes in the region to be served, degree of compaction imparted to the wastes, number of lifts, and depth of temporary (if more than one lift) and of the face earth covers. Because the pertinent rate of generation of wastes is that of the population for which the fill is intended, rate then becomes a function of size of population and amount of wastes generated per person per day plus the daily output from commercial sections and other designated contributors. Generally rates of generation are expressed gravimetrically rather than volumetrically, mostly because gravimetric measurements are more readily obtained. Truckloads of material can be weighed at the disposal site, and the weights then can be extrapolated to volume by applying conversion factors reported in the literature. Curves indicating the relation between amounts of solid waste collected or disposed per day and the volume of those wastes at various degrees of compaction are shown in Figure 2. A detailed account of the steps involved in determining site capacity are clearly explained by Cavanagh.[16]

## VII. USE OF COMPLETED FILL

Eventually the capacity of the sanitary landfill is reached. If it had not been done as the fill progressed, the permament cover is applied, and other modifications in keeping with the intended use of the fill are made. These tasks having been finished, the fill conventionally is referred to as a "completed fill".

Several important factors place a sharp limitation on the uses to which a completed fill can be put. Among them are bearing capacity of the soil, settling, and the presence of methane and of corrosive substances. The bearing capacity of a completed fill is low and varies with the manner in which the fill was made. Values of 2443 to 2906 kg/m$^2$ (500 to 800 lb/ft$^2$) have been reported. Because of the low bearing capacity, only light-weight structures should be errected on a completed fill. To support a heavy structure it would be necessary to rely upon piling.

Settling is an extremely important limiting factor for obvious reasons. It is a result of the further increase in density that results from the gradual rearrangement of the particles (sifting action) and from the weight of the upper strata of wastes and of the cover material, as well as of the decomposition of the wastes. Because of the first factor, some settling is to be expected regardless of the nature of the wastes. However, the major share of settling depends upon the nature of the wastes in the fill, the extent to which they decompose and the degree of compaction imparted to them, as well as on the depth of the fill. Certainly, the larger the organic fraction and the deeper the fill, the greater will be the extent of settling. Since rate of settling is a function of that of decomposition, it also is a function of the factors that affect decomposition. Usually about 90% of the total settling takes place during the first year of the fill. An example of the dimensions of settling and the effect of decomposition follows. In Seattle where the annual rainfall exceeds 11 cm/yr, a 6-m fill settled 1.2 m in year 1; in Los Angeles County, where the rainfall averages less than 6 cm/yr, a 23-m fill had settled only 0.7 m by year 3, and a 14-m fill, only 0.4 m.[4]

The presence of methane is important, because as explained earlier, it is a potentially explosive gas. The greatest danger in this respect comes from the accumulation of the gas in substructures of buildings erected on the fill.[17] The mere presence of an asphalt cover over the fill does not preclude the diffusion of methane under a building, because the gas can seep through cracks in the asphalt. The danger of methane accumulation also extends to nearby structures because of the possible lateral migration of the gas. The corrosive materials are the organic acids formed during biological decomposition. Their effect would be on utilities buried in the fill. Consequently, all utilities must be appropriately protected from the action of the acids.

In view of the limitations, perhaps the best use for a completed fill would be as an *open* or *green* area, e.g., park or garden.

Because the restrictions on the use of the completed fill lasts for many years, it has been recommended that a notice of the existence of the fill be incorporated into the land deed so that future buyers may become cognizant of the limitation.

## VIII. PROCESSING WASTES FOR LANDFILLING

The two principal developments in recent years for processing wastes for landfill are size reduction and baling. The rationale for size reduction, according to the proponents of the practice, is as follows: (1) the capacity of the fill is increased because of the increase in compressibility of the wastes from milling; (2) decomposition is hastened; and (3) the need for a cover is eliminated. Claims range from a 30% to a negligible reduction in volume of the wastes. The reasoning is that the smaller the particle size, the greater is the potential degree of compression. Whether or not the goal of increased landfill capacity is reached remains open to question. According to results reported by Ruf,[18] reduction in particle size does have a positive effect on reduction of bulk density within the pressure range of 69 to about 1103 to 1241 $kN/m^2$, but the difference becomes much less at higher pressures because of rebound. Size reduction does substantially increase the rate of decomposition, which in turn leads to the production of a stronger leachate initially. However, the reduction in leachate strength is correspondingly more rapid.[19] Conceivably the accelerated decomposition could be attended by a consequent increase in initial settling rate. The claim of eliminating the need for cover material is subject to dispute. Existing guidelines call for a cover, albeit of a somewhat reduced thickness. The deciding factor, assuming all of the claims of the proponents to be true, is the trade-off between the added cost of size reducing the wastes (about $3.30/Mg) and the amount of the savings resulting therefrom.

Bales produced for landfill generally are 1.2 to 1.8 m long and 0.9 to 1.2 m in cross section, and have a density of 971 to 1132 $kg/m^3$. In the fill they are placed in layers two or three bales high per lift. Compaction is needed only for the daily cover soil. According to the proponents, the advantages accruing from the incorporation of baling into a sanitary landfill operation are (1) most of the objectionable sanitary and aesthetic objections associated with refuse are minimized; (2) the compaction effort at the landfill is eliminated, as is the subsequent settlement; and (3) earth cover requirements are lessened. The lessening of the compaction effort makes possible a drastic reduction in the equipment needs for a fill. The bulldozer, scraper, compactor, etc. are replaced by a forklift or a front-end loader. Even if all the claims were to materialize, the deciding factor remains one of comparative economics. In terms of costs, baling entails an expenditure of about $1.10/Mg (1974 dollars) over and above that of the landfill operating costs.

With both size reduction and baling, it should be kept in mind that the cost estimates are those offered by proponents. If experience in general is an indicator, the actual costs are apt to be much higher.

## IX. COSTS

Because of the certainty of the continuation of the inflationary spiral and the uncertainty as to its rate of increase, the values named herein are given mainly to indicate relative magnitudes and the relation of the cost of one item in comparison to that of another.

Exclusive of land, the most expensive of the capital costs is that of equipment. Total equipment cost naturally increases with the size of the operation. Equipment requirements for various sizes of operations are listed in Table 5, and the costs (1970) of the machines in Table 6. With respect to operation, expenditure on fuel accounts for about 90% of the costs. The remaining 10% is spent on labor, maintenance, utilities, and miscellaneous necessities.

**Table 5**
**EQUIPMENT REQUIREMENTS FOR A SANITARY LANDFILL OPERATION**

| Solid waste handled (Mg/8 hr) | Crawler loader | | Crawler dozer | | Rubber-tired loader | |
|---|---|---|---|---|---|---|
| | Flywheel (kW) | Weight[a] (Mg) | Flywheel (kW) | Weight (Mg) | Flywheel (kW) | Weight[a] (Mg) |
| 0—18 | <52 | <9.0 | <60 | <6.8 | <74 | <9.0 |
| 18—41 | 52—74 | 9.0—11.3 | 60—82 | 6.8—9.0 | 74—89 | 9.0—10.1 |
| 41—118 | 74—97 | 11.3—14.6 | 82—97 | 9.0—11.3 | 89—112 | 10.1—12.4 |
| 118—227 | 112—142 | 14.6—20.3 | 112—134 | 13.5—15.8 | 112—142 | 12.4—15.8 |
| 227—454 | Combination of machines | | 186—209 | 21.4—23.4 | Combination of machines | |
| 454—plus | Combination of machines | | | | | |

[a] Basic weight without bucket, blade, or other accessories.

Modified from Brunner, D. R. and Keller, D. J., Sanitary Landfill Design and Operation, (SW-65ts) U.S. EPA, 1972.

**Table 6**
**EQUIPMENT CAPITAL COSTS**

| Machine type | Flywheel (kW) | Weight (Mg) | Approximate weight[a] (Mg) | Approximate cost[b] ($) | Comments |
|---|---|---|---|---|---|
| Crawler dozer | <60 | <6.8 | 8.6 | 9.5 | Landfill blade |
| | 82—97 | 9.0—11.3 | 14.4 | 17.1 | Landfill blade |
| | 186—209 | 21.4—23.4 | 30.2 | 31.5 | Landfill blade |
| Crawler loader | <52 | <9.0 | 9.5 | 9.5 | GPB[c]: 0.8 $m^3$ |
| | 75—97 | 11.3—19.6 | 14.0 | 13.5 | GPB: 1.5 $m^3$ |
| | 75—97 | 11.3—14.6 | 14.4 | 14.4 | MPB[d]: 1.3 $m^3$ |
| | 119—142 | 14.6—20.1 | 20.3 | 20.7 | GPB: 2.3 $m^3$ |
| | 119—142 | 14.6—20.1 | 21.2 | 22.1 | MPB: 1.9 $m^3$ |
| Rubber-tired loader | <75 | 9.0 | 7.7 | 9.5 | GPB: 1.3 $m^3$ |
| | <75 | 9.0 | 8.1 | 9.5 | MPG: 1.1 $m^3$ |
| | 89—119 | 10.1—12.04 | 9.5 | 14.9 | GPB: 3.0 $m^3$ |
| | 89—119 | 10.1—12.4 | 11.7 | 16.2 | MPB: 1.7 $m^3$ |

[a] Basic machine plus engine sidescreens, radiator guards, reversible fan, roll bar, and either a landfill blade, general-purpose bucket, or multiple-purpose bucket as noted.
[b] June 1970.
[c] General-purpose bucket.
[d] Multiple-purpose bucket.

Modified from Brunner, D. R. and Keller, D. J., Sanitary Landfill Design and Operation, (SW-65ts) U.S. EPA, 1972.

The actual operating and maintenance expenses can be determined when the site is in operation through the use of a cost accounting system such as that described by Zausner.[20] Excluding the cost for land, that total cost for landfilling a metric ton (Mg) of municipal solid waste in 1979 ranged from $3.00 to $5.00.

## X. LANDFILL GAS COLLECTION

A characteristic of landfills that in the past was regarded as a handicap is the generation of methane that normally takes place in one. With the realization of the existence of an energy shortage, the attitude towards such methane generation is changing from one of regarding it as a handicap to one of viewing it as an advantage, i.e., a recoverable resource. An outcome of the change is that since 1975, several projects have been proposed and initiated, all of which involve methane recovery and utilization either from existing fills or from fills as yet underway. With the latter, the design of the fill is altered hopefully to provide for maximum recovery.

### A. Volume and Composition of Gas

Various methods can be followed for estimating the potential yield of methane from a fill. However, it should be borne in mind that with all of the methods, the resulting estimates are far from firm because of the mostly uncontrolled conditions encountered in a fill. Consequently, it would be sheer optimism to expect that all organic matter in the buried wastes will be decomposed by the methane fermentation route. Moreover, a sizeable portion of the carbon becomes carbon dioxide as a result of microbial metabolism, although a minute fraction of the $CO_2$ may be used as a carbon source by methanogens. Moreover, varying amounts of the methane actually generated are lost from the fill by diffusion to the surrounding environment. With these strictures in mind, the method described in the section "Methane Fermentation" Volume II, chapter 4 is cited as one of the more "optimistic" methods. It is a result of experience gained in the digestion of sewage sludge and relies upon amount of volatile matter as a basis for calculations. According to the method, the maximum yield of gas would be on the order of 0.374 $m^3/kg$ volatile (organic) solids in the buried wastes. However, a more likely yield would be within the range of 0.013 to 0.047 $m^3/kg$.[21,22] These rates prevail during the period of most active methane formation, which usually is during the first 5 years or so after landfilling.[23] The composition of the gas varies with the composition of the wastes and the age of the fill. In general, it will be on the order of 40 to 50% $CH_4$, 30 to 40% $CO_2$, 10 to 20% $N_2$, 1% $O_2$, and traces of sulfides.

### B. Gas Collection Systems

If methane recovery is to be a part of a new sanitary landfill, certain features must be incorporated into the design of the fill. Some are characteristic of landfill design even where methane recovery is not a goal, while others are peculiar to recovery. A general requirement is that the fill be effectively sealed off its land and water environment, and relatively so from the air above it. A feature more or less peculiar to gas recovery is the provision of a gas collecting system consisting of strategically spaced collection wells. The collection wells are gravel-packed wells equipped with casings that penetrate the full depth of the fill and are perforated in the section exposed to the fill contents. Each casing must have telescopic connections between the pipe segments to maintain continuity in it despite the signficant and nonuniform subsidence characteristic of landfills. A diagrammatic sketch of one such well is shown in Figure 7.

The spacing of the gas collection wells depends upon their capacity as well as upon the characteristics of the soil cover. The radius of the influence of a gas collection well is a

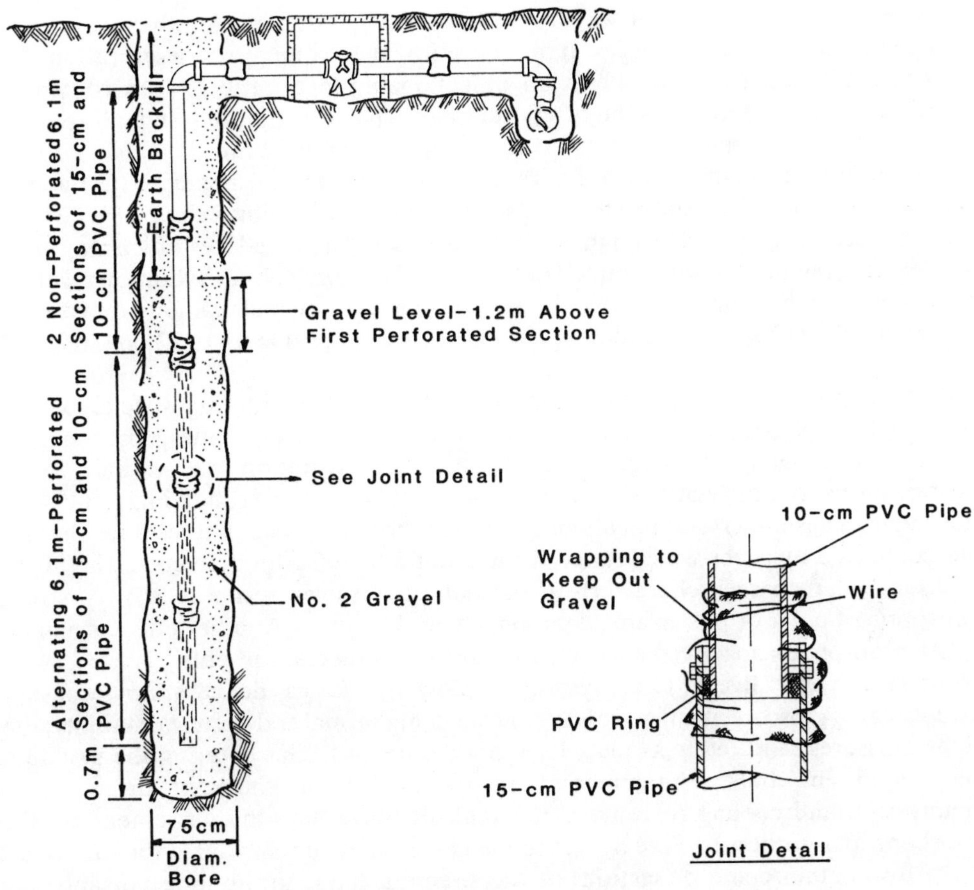


FIGURE 7. Gas collection well.

function of the rate of pumping. For example, in a 12-m California fill with a 6-m deep gas well, and the pumping rate at 2.83 $m^3$/min, the negative pressure in the fill ranged from −5.1 cm of water at the well to less than −0.8 cm at a distance of 30.5 m from the well. When the rate was advanced to 8.5 $m^3$/min, the respective negative pressures became −17.8 and −2.54 cm.

The rate of withdrawal of the gas must not be so great as to pull air through the cover and into the fill. This precaution is especially important if the cover material is relatively porous. In the Los Angeles experience, the optimum rate of withdrawal proved to be 1.132 $m^3$/min.[24] The undesirable effects of drawing air into the fill are twofold: (1) the concentration of $N_2$ in the biogas is increased, thus lowering its quality; and (2) oxygen in the air thus introduced inhibits the activity of the methane-generating microorganisms.

### C. Upgrading and Utilizing the Gas

Excepting for simple space heating and household cooking, biogas should be upgraded before being put to use. This is especially important if the gas is to serve as a fuel for an internal combustion engine, and if the gas is to be injected into existing transmission lines, the need is even more imperative. The quality, content, and day-to-day consistency of landfill gas differ by far from those of natural gas. Among the undesirable differences are a low heat content of from 7450 to 22,350 kJ/$m^3$, as compared to that of natural gas, which at its lowest is 37,260 kJ/m$^3$. Another drawback is the moisture content of landfill gas. It may be as low as 5% or may reach saturation. The hydrogen sulfide content can be at an objectionable level. The oxygen content of landfill gas varies widely. If too high, a potentially explosive mixture results. Variations in the preceding parameters can be as much as 50% from day to day and season to season. Other detracting characteristics are a sizeable carbon dioxide ($CO_2$) and $N_2$ content, both of which contribute to a lowering of the heating value.

The gas can be upgraded through one or more of several procedures which may range in extent from simple dehydration to $H_2O$, $CO_2$, and $N_2$ removal. Dehydration results in an increase of heating value by about 10% of the original value. Dehydration accompanied by $CO_2$ and $H_2S$ removal brings the heating value up to 22,360 to 26,000 kJ/m$^3$.

Methods of removing water include in-line gravity outflow, filtering, special solvents (polyethylene glycol), molecular sieves, permaselective membranes,[25] heating, air cooling, and refrigerant cooling. Of this collection of dehydration systems, the one that involves the use of a molecular sieve combines a relatively low cost and a high efficiency. Molecular sieves are crystalline aluminosilicates, honeycombed with cavities that are interconnected by pores which range from about 3 to 100 Å in diameter. The highly localized polar charges characteristic of molecular sieves result in a very strong absorption of polar or polarizable compounds on the sieves. Another effect is a much higher absorptive capacity than is characteristic in other absorbents.

The Triethylene Glycol (TEG) system is widely used for gas dehydration because of the unusual hygroscopicity of glycols, their excellent thermal and chemical stability, low vapor pressures, and ready availability at moderate cost. Gas entering the system is compressed and bulk contaminants are removed in a knockout drum. After compression and cooling to remove the bulk of the water, the gas enters a TEG absorber/separator tower. Free liquids in the gas stream are removed in the lower part of the tower (the separator section) as the stream ascends to the upper or absorber section of the tower. In the absorber section, the gas stream contacts lean triethylene glycol on bubble-cap trays.

Dehydration and $CO_2$ and $H_2S$ removal can be accomplished simultaneously by coupling the TEG dehydration system with a hot potassium carbonate scrubbing system.

If the utilization of the gas requires the removal of hydrogen sulfide only, passing the gas through a dry gas scrubber containing an "iron sponge" made up of ferric oxide mixed with wood shavings is an excellent approach.[24] One cubic meter of iron sponge will remove 105 kg of sulfur. Regeneration of the "sponge" is accomplished by exposing the iron sponge to air. The ferric sulfide formed in the scrubbing operation is converted to ferric oxide and elemental sulfur.

The upgraded gas can be used on-site to generate electricity, or may be injected into a public utility transmission line. Three possiblities exist for on-site electrical generation. They are the use of the fuel in an internal combustion engine, the driving of a gas turbine, or a fuel cell. If an internal combustion engine (diesel engine) is involved, the landfill gas is compressed to about 5 psig. If a gas turbine is used, the pressure is increased to 150 psig.

## D. Costs

Although costs will vary from installation to installation, and certainly with time because of the inflation factor, a few overall numbers are presented to indicate an idea of relative costs in terms of other energy sources. According to information furnished by Pacific Gas and Electric (northern California), it would cost about $1.30 (1976 dollars) per kJ to dehydrate and compress (420 psig) raw gas at 28,300 $m^3$/day. The heating value of the gas would be 16,767 kJ/kg, and the moisture content, 7%. At the same daily throughput, the cost to deyhdrate the raw gas to 7% moisture and remove its $CO_2$ and $H_2S$, and to compress it to 420 psig, would be $2.17 per million kJ. The heating value would be 22,356 to 26,080 kJ/$m^3$. A stream factor of 0.85, 10-year landfill life span, 20-year equipment life, and a 12% rate of return are assumed. Of the costs, that of compression would account for at least 50%.

**E. Decision Factors**

While the prospect of obtaining a useful fuel gas by way of a seemingly terminal waste disposal method may at first sight appear to be an attractive one, in certain cases several factors may combine to dim the prospect considerably in terms of economic feasibility and of maintaining a positive energy balance. Among the more important factors are size and location of the fill, permeability of cover material and surrounding soil layer, climatic conditions, and proposed use of the gas. While appropriate designing can provide for an impermeable surrounding of the fill, size and location often are not subject to alteration, and certainly, little can be done about climatic conditions. As far as use of gas is concerned, if the use requires a top quality gas, then the cost of upgrading the biogas may be prohibitively high.

Obviously, the contents of the fill should be sufficiently great as to ensure an eventual total gas output, the monetary and energy value of which would surpass that expended in the departures from conventional fill practice needed to recover the gas. Not to be neglected in arriving at a suitable estimate is the relatively brief period of active gas production. This limitation would preclude expensive equipment, unless the equipment were portable. Climatic conditions affect requisite size in that an unfavorable set of conditions would slow the rate of gas generation, and hence the permissible rate of withdrawal. An exception to the latter could be made if the gas were not withdrawn until the completion of the entire fill. The reduction in rate of withdrawal would expand the landfill volume required to render the project feasible.

At this time, available information warrants only the generality that the sizes of the majority of landfills in the nation are not sufficiently large to make biogas recovery a worthwhile undertaking.

With respect to dimensions, it would be inadvisable to utilize a fill that is less than 13-m deep. A detailed discussion on size and economics is presented in the EPA report "Recovering, Processing, and Utilization of Gas from Sanitary Landfills."[23] According to the report, to have a favorable prognosis for economic feasibility, the completed fill should contain at least about two million tons of municipal solid waste. At the peak rate of generation, the gas produced from a fill of such a dimension would be from 28 to 34 $m^3$/min of raw gas. The equivalent in recovered energy would be 759 mmkJ/day.[23]

The significance of permeability certainly requires no explanation. The more impermeable the surrounding soil layer, including the cover, the greater will be the loss of methane from the fill. If a fill were constructed and operated in conformity with the EPA sanitary landfill guidelines, permeability would not be a serious problem.

The proposed use of the gas has a decisive bearing both on economics and energetics. As indicated in a preceding paragraph, upgrading biogas to pipeline quality is an expensive operation. The present consensus is that such a procedure would not be economically viable for a private enterprise. As far as local governments are concerned, it might be done on a break-even basis with the larger landfills if the gas were sold at wholesale prices to a local utility. Probably, the safest use in terms of economics is as a fuel to be used in steam generation, at the landfill site, or for an internal combustion engine after a minimum of cleaning. On-site use is recommended because the moisture and the presence of corrosive elements in the raw biogas militate against pipeline transport.

## REFERENCES


1. **Stearns, R. P. and Woodward, J. P.,** Resource recovery and the need for sanitary landfills, *Publ. Works,* 108(9), 160, 1977.
2. **Greco, J. R.,** How some states are defining sanitary landfills, *Solid Wastes Manage./Refuse Removal J.,* 20(4), 26, 1977.
3. Anon., A compactor but hardly a compact, *Waste Age,* 8(12), 24, 1977.
4. **Brunner, D. R. and Keller, D. J.,** Sanitary Landfill Design and Operation, (SW-65ts) U.S. EPA, 1972.
5. **Lutton, R. J., Regan, G. L., and Jones, L. W.,** Design and Construction of Covers for Solid Waste Landfills, EPA-600/2-79-165, Municipal Environmental Research Laboratory, Office of Research and Development, U.S. EPA, Cincinnati, Ohio, August 1979.
6. **Stoeng, D. R.** The effects of the disposal of industrial waste within a sanitary landfill environment, unpublished paper, Systems Technology Corp., Cincinnati, Ohio, 1978.
7. **Fenn, D. G, Hanley, K. J. and DeGeare, T. V.,** Use of the Water Balance Method for Predicting Leachate Generation from Waste Disposal Sites, (SW-168) U.S. EPA, 1975.
8. **Salvato, J. A., Wilke, W. G., and Mead, B. E.,** Sanitary landfill-leaching prevention and control, *J. Water Pollut. Control Fed.,* 43(10), 2084, 1971.
9. **Armentrout, W. F., Jr. and Bortner, L. L.** Preliminary Investigation on Landfill Stabilization with Leachate Recycle, special research problem, Georgia Institute Technology, School of Civil Engineering, Atlanta, Ga., September 1971, 55.
10. **Pohland, F. G.,** Sanitary Landfill with Leachate Recycle and Residual Treatment, EPA—600/2 75-043, U.S. EPA, October 1975, 105.
11. **Fuller, W. H. and Korte, N.,** Attenuation Mechanisms of Pollutants through Soils, in Gas and Leachate from Landfills, Proc. Res. Symp. Rutgers University, New Brunswick, N.J., EPA-600/9-76-004, U.S. EPA, March 1975.
12. **Griffin, R. A. and Shimp, N. F.,** Leachate Migration through Selected Clays, in Gas and Leachate from Landfills, Proc. Res. Symp. Rutgers University, New Brunswick, N.J., EPA-600/9-76-004, U.S. EPA, March 1975.
13. **Farquhar, G. J. and Rovers, F. A.,** Leachate Attenuation in Undisturbed and Remoulded Soils, in Gas and Leachate from Landfills, Proc. Res. Symp. Rutgers University, New Brunswick, N.J., EPA-600/9-76-004, U.S. EPA, March 1975.
14. **Moore, C. S.,** Theoretical Approach to Gas Movement through Soils, in Gas and Leachate from Landfills, Proc. Res. Symp. Rutgers University, New Brunswick, N.J., EPA-600/9-75-004, U.S. EPA, March 1975.
15. **Flower, F. B.,** Case History of Landfill Gas Movement through Soils, in Gas and Leachate from Landfills, Proc. Res. Symp. Rutgers University, New Brunswick, N.J., EPA-600/9-76-004, U.S. EPA, March 1975.
16. **Cavanagh, T. E., Jr.,** How to collect preliminary data for location, design, and authorization of a sanitary landfill, *Solid Wastes Manage./Refuse Removal J.,* 18(4), 22, 1975.
17. Development of Construction and Use Criteria for Sanitary Landfills, interim report prepared by the County of Los Angeles, Department of County Engineering, Los Angeles and Engineering Science Inc. or the U.S. Department of Health, Education, and Welfare, 1969.
18. **Ruf, J.,** Particle Size Spectrum and Compressibility of Raw and Shredded Municipal Solid Waste, Ph.D. dissertation, University of Florida, Environmental Sciences, 1974.
19. **Dusynski, E. J.,** Grinding refuse is no experiment, *Am. City,* 86(9), 61, 1971.
20. **Zausner, E. R.,** An Accounting System for Solid Waste Management in Small Communities, Report SW-28ts, U.S. EPA, 1971.
21. **Bowerman, F. R., Rohatgi, N. K., Chen, K. Y. and Lockwood, R. A.,** A Case Study of the Los Angeles County Palos Verde Landfill Gas Development Project, EPA-600/3-77-047, Municipal Environmental Research Lab, U.S. EPA, Cincinnati, Ohio, July 1977 (U.S. Dept. Commerce NTIS PB-272 241).

22. **Hekimian, K. K., Lockman, W. J. and Hirt, J. H.,** Methane gas recovery from sanitary landfills, *Waste Age,* 7(12), 2, 1976.
23. **Ham, R. K., Hekimian, K. K., Katten, S. C., Lockman, W. L., Lofy, R. J., McFaddin, D. E. and Daley, E. J.,** Recovery, Processing, and Utilization of Gas from Sanitary Landfills, EPA-600/2-79-001, Municipal Environmental Research Lab, U.S. EPA, Cincinnati, Ohio, February 1979.
24. **Blanchet, M. J.,** Production, Treatment and Utilization of Landfill Gas, report prepared for the U.S. EPA, Shoreline Regional Park Gas Recovery Program, by the Pacific Gas and Electric Co., San Francisco, Calif., 1976.
25. **Walmet, G. E. and Boyd, W. D.,** Biogas purification with permaselective membranes, in *Biogas and Alcohol Fuels Production,* Proc. Semin. Biogas Energy for City, Farm, and Industry, Chicago, Ill. May 1979, J. G. Press, Emmaus, Pa., 1979.

# INDEX

## A



## B

## D

## E



## F

## I



## J



## K



## L



## M

## N



## O

## P

## T



## U



## V



## W

**Y**



**Z**